AF412549

Quality Characterization of Apparel

2nd edition

Quality Characterization of Apparel

2nd edition

Dr. Subrata Das

WOODHEAD PUBLISHING INDIA PVT LTD

New Delhi, India

Published by Woodhead Publishing India Pvt. Ltd.
Woodhead Publishing India Pvt. Ltd.,
303, Vardaan House, 7/28, Ansari Road,
Daryaganj, New Delhi - 110002, India
www.woodheadpublishingindia.com

First published 2019, Woodhead Publishing India Pvt. Ltd.
© Woodhead Publishing India Pvt. Ltd., 2019

Woodhead Publishing India Pvt. Ltd. ISBN: 978-93-88320-10-8
Woodhead Publishing India Pvt. Ltd. E-ISBN: 978-93-88320-11-5

Typeset by Versatile PreMedia Services (www.versatilepremedia.com)

This page is intentionally left blank

This page is intentionally left blank

Contents

This book arose out of a need, when interacting with the apparel fecundity at different parts of the world, for a comprehensive guideline on the quality of various merchandise products to which different stakeholders could be referred. The approach to the subject and the topics covered are those, which have been developed over the years on the global platform in apparel sectors by retailers and regulatory bodies. The apparel industry has had a long history of producing different merchandise products in accordance with the necessity of various classes of consumers and the nature of their utilisation. Due to this various quality parameters have been evaluated and benchmarked by following international standard norms to match the performance requirements of different apparel products. This book is then a distillation of these collective efforts and hopefully a concise document of wisdom inculcated over the years in the evaluation of quality of apparel.

The book is aimed at textile and apparel industry professionals, retailers, factory heads, buying offices, and students intending to join the industry in the areas of quality assurance. In order to produce and deliver better quality products to the customer, adherence to the appropriate specification, standard, law, and regulation applicable for the merchandise is important. Thus, the emphasis throughout the book is on standard and mandatory regulatory test methods. Many of the apparel merchandise benchmarked are intended to evaluate the same property, but specification and regulation may vary because of their different export destinations. It is worthwhile to mention that when referring to any tests and regulations to consult an up-to-date version of the relevant document. This is because the actual standard contains the detail information, which is not possible to cover in a book of this nature, and furthermore international standards and regulations are constantly being revised and updated due to the change of performance expectations and better consumer protection.

Safety of children's apparel has been given more importance in quality during recent years by the apparel retailers in the globe, but is the least discussed subject available in the published literature. The commitment towards this important aspect of quality is expected from the apparel manufacturers to supply and meet the requirement of global retailers. Failure to focus on the necessary safety issues results in product recall. Thus, the book covers about the use of different accessories, which can be attached in children's apparel with the associated safety review product evaluation and, regulatory approaches.

The first edition of the book was published in the year 2009. Since then lot of changes happened in the apparel industry. Keeping in view of the recent fashion trend and the feedback received from the global readers, two new chapters have been added to the second edition of the book. Revision has also been incorporated in the existing chapters. In recent days, denim fabrics are the most favourite choices for the fashion world to produce innovative products. The reason being different washing treatments to develop various aesthetic look and fantastic feel. But such a treatment changes physical and chemical characters of the original denim fabric. So, in the newly introduced chapter, varieties of washed denim fabrics are characterized to understand changes happened to their properties. Apart from denim fabrics, use of different types of leather hides are gaining popularity in the apparel sector along with special finishes to make the product enchanting to the end users. Thus, quality characterization of different finished leather and performance requirements of different leather and suede garments have been added to the second edition of the book for a wider coverage of the area of quality characterization of apparels.

Date : 31st May 2018
Place : Sathyamangalam

Dr. Subrata Das
Professor (Fashion Technology)
Bannari Amman Institute of Technology,
Sathyamangalam, Erode District,
Tamil Nadu - 638401
India

CHAPTER 1

Introduction

Abstract

Global business in apparel sector is dependent on quality characterisation because major buyers want to ensure about the quality of the merchandise prior to the delivery to the consumers. The chapter first discusses the importance of product quality, which is dependent on fibre and fabric type, weight, style, finish, accessories used, country of export, and above all the intended end use. The chapter then discusses about the essential elements, which are to be addressed as per international standard norms and the role of regulatory, and specialty tests to prevent recalls and to enhance profitability.

Keywords: quality; testing protocol; drycleaning; performance; safety

Chapter contains (Section headings)

1.1 Importance of quality characterisation
1.2 Current scenario
1.3 Essential elements of quality characterisation
    1.3.1 Dimensional properties
    1.3.2 Colour fastness properties
    1.3.3 Durability and surface appearance
1.4 Role of regulatory and specialty tests in quality characterisation
1.5 Customer satisfaction related to quality
References

1.1 Importance of quality characterisation

In the apparel sector, quality control is practiced right from the initial stage of sourcing raw materials to the stage of finished garment. Product quality is assured in terms of fibres, yarns, fabric construction, colour fastness, durability, surface designs, garment construction, and the final finished item. However, quality expectations for export are related to the type of customer segments and the retail outlets. In today's competitive business of apparel export, characterisation of quality is an important and indispensable aspect. Global standards in apparel are technology driven, benchmarked by the major buyers and ultimately product oriented.[1] Tolerances in the degree of product proficiency cannot be ignored since too slack standards may allow excessively inferior merchandise to pass through, while, standards, which are too rigid, may result in acceptable merchandise being unnecessarily rejected. Thus, quality evaluation of garments as per international standard norms is essential for export. This is not only to ensure a quality product, but also to endorse the product safety, prevent recalls, reduce returns, minimize customer complaints, and promote repeat sales. It is well known that testing protocols are the summaries of applicable requirements, which cover all facets of performance, evaluating safety and quality, as well as labelled claims. Due to ever increasing fashion trend, different fibre, construction, style, colour, and finish dominate the apparel world to cater to the requirements of various categories of customers. But, unfortunately, no single universal characterisation protocol is available in the garment trade. Testing protocol changes depending on the fibre and fabric type, weight, style, finish, accessories used, country of export, and above all the intended end use of the product. It is also vital to bear in mind that all standards and regulations encapsulated in the protocol have one or both of the following aims: safety and quality. While quality is related more towards general consumer satisfaction, safety is an important concern as products not meeting regulations can jeopardize the health of the purchaser. Thus, characterisation of apparels that are earmarked for export is essential to satisfy both the regulation and requirement. Any deviation in production and quality goes against the interest of consumer who is the ultimate user. As a result, the brand image gets affected due to poor presentation and performance of an apparel product under question.

1.2 Current scenario

Apparel trade in post-quota regime has transformed the business world to a global village. The old concept derived from four P's: product, price, place, and promotion has been replaced with a new pattern, the four C's: consumer, cost, convenience, and communication.[2] Undoubtedly, there is more competition on a level playing field since dependence on quota-profile is no more be an advantage of any country. To survive in this network, there is a paradigm shift towards attitude in working out strategies in the garment arena. Assurance of international standards, product innovation and adaptability to changing tastes of consumers are some of the areas of current interest. To win over apparel consumers, manufacturers, brands, and retailers are struggling hard to identify "product value". Undoubtedly, fashion sells, but only to a point, beyond which many consumers are searching values in apparel. Traditionally, "value" is been defined as a function of price and quality.[3] There is a section of this niche market that is price conscious, but largely this section is also brand aware and would not mind to spend more to buy good branded apparel. Thus, today's consumers are redefining value to include reliability of the product performance—they are asking whether product is assured in actual use or not.

There are some of the inevitable questions, which need to be addressed in the real life situation on garment performance in actual use: Does the garment shrink? Will it loose colour? What about its durability? Will the garment torque? Is there any harmful substances in the garment?

1.3   Essential elements of quality characterisation

Aesthetics in apparel are not expected to sacrifice for durability and performance. Thus, when consumers buy apparels they suspect about the quality and expect some change in shape and colour after washing. But the degree, to which this happens, entails the difference between satisfaction and disappointment. Prediction of such performance is only possible through comprehensive quality evaluation. Many garment properties are important to the final customer. Some are highly specialised in nature, but there is a core series of tests that are applicable depending on the end use of the product. There exist internationally recognised standards applicable for Europe and United States and broadly denoted as ISO, BS, EN, BSEN, DIN, ASTM, and AATCC. In addition, many retailers around the world have their own standards and test methods, e.g., Marks and Spencer, J.C. Penny etc. Methodology and equipment vary but basic objectives remain the same.

1.3.1   Dimensional properties

A common concern in apparel performance characterisation is the dimensional stability. Accelerated test methods[4,5] are applied to wash and dry at the recommended conditions and careful measurement of any changes in dimensions determines the product ability to withstand the "care label" recommendations. Side seam twisting or garment torque[6] for knitted goods is the most common problem, which can be quantified by appropriate test method. Controversy arises while marking and taking measurement. Proper understanding of the test method, appropriate application and right interpretation of the test result are important to assure the desired characterisation.

1.3.2 Colour fastness properties

Apparel products fade due to various actions. It can be a particular problem with lower cost materials and processes, where insufficient care is been taken during dyeing, or sometimes because of the limitations of technology. In general, the tests measure the degree to which the colour changes when treated in a way that simulates the conditions of use such as washing,[7] drycleaning,[8] water,[9] perspiration,[10] rubbing,[11] chlorine, and non-chlorine bleach,[12] chlorinated water,[13] light,[14] gas fume fading,[15] ozone fading,[16] and print durability. Many tests also measure the degree of colour transfer on uncoloured fibres in the same environment. Keeping in view of the application of various linings and varieties of different fibre panel in the same apparel, the use of fibre types as per standard, namely acetate, cotton, nylon, polyester, acrylic, and wool is an essential part in certain tests to judge staining behaviours. Requirement of appropriate testing method and its proper application are of paramount importance in arriving at a conclusion of suitability of apparel intended for a specified end use. Conditions require for testing fashion apparel fading when exposed to light being different with respect to testing the fading of upholstery in a car interior. These different end users need to be accounted for during quality evaluation, even though the fading of either product under their normal conditions of use will cause a problem.

1.3.3 Durability and surface appearance

Durability of a garment is another important criterion to predict its permanence in use. Consumer often enquires this fact before final selection of the apparel product. It is basically the power of a garment to resist stress or force. Test procedure typically subjects the material to stress of some kind, and measures the amount of force at which it fails. Most common tests measure the tensile,[17] tear,[18] and bursting strength.[19] Deterioration of surface appearance due to abrasion and pilling are other areas wherein customer dissatisfaction is quite common. This cannot be ignored since often garments develop unsightly ball of fibre, which are noticed to the surface of the fabric. Sometimes those are of a slightly different colour from the main fabric, and can ruin the wearability of the article. Pilling tests[20] provide the valuable guidance of the pilling performance of fabrics.

1.4 Role of regulatory and specialty tests in quality characterisation

Customers of 21st Century are also safety and health conscious. Thus, it is well accepted that they will not mind to pay extra for this cause. Specialised quality characterisation in apparel includes flammability, fibre composition, restricted harmful substances, and performance tests such as water and oil repellence, stain resistance, bacterial resistance, or breathability, yellowing in storage etc. Out of which, fibre composition test is more of a protective regulation against dishonesty. Consumer paying a high price for a silk garment would be most unhappy to discover that it was made of polyester. Protection of the consumer interest surely supplements by determining the fibre composition as per international standard methods of testing. The fibre content label required by federal law in US to be permanently attached in each garment must indicate the percentage of each fibre present in the garment. Regulatory Consumer product testing of flammability and restricted substances in adult and children apparel are important from the point of view of safety and health. If risk involved in wearing of a garment is known beforehand by flammability performance testing or clothing is qualitatively and quantitatively analysed for potentially harmful substances, such as formaldehyde, harmful azo dyes or carcinogenic and allergenic dyes, heavy metal content, phthalates, PCP or TeCP and organotin compounds, it definitely protects the interest of a consumer belongs to different segments. Testing for harmful substances plays a significant role in the considerations made when buying textiles. This is demonstrated by the results of a trend analysis survey commissioned by the German OEKOTEX® Certification Centre and carried out by the consulting firm BBE Retail Experts in the Netherlands, Austria, Switzerland, Portugal, Italy, France, and Spain.[21] According to the opinions of the specialist retailers surveyed, product quality, social aspects, skin compatibility, and testing for harmful substances were the most important parameters for customers when buying textiles and were regarded correspondingly by the retailers in their ordering behaviour. On a scale from 1 to 5, with 1 = unimportant to 5 = very important, these factors constantly received ratings ranging between 4.2 and 4.6. When asked about the significance of testing for harmful substances, the rating of 4.2 from Germany was consistent with the average of the ratings of the seven other European countries surveyed.

1.5 Customer satisfaction related to quality

In the apparel sector, it is well perceived that quality is a multi-dimensional aspect. There are many areas of quality based on which the garment exporters are supposed to work. Quality of the production, quality of the design of the garment, quality of purchase, quality of final inspection, quality of the sales, and quality of marketing of the final product are some of the important measures. But quality of the final product is ultimately integrated to customer satisfaction. Quality increases the value of a product or service, establishes brand name, and builds up reputation for the garment exporter, which in turn results to build consumer confidence, high sales, and foreign exchange for the country.²² It is worthwhile to mention that the responsibility of production of appropriate quality garment and right characterisation as per international standard norms lies with different agencies such as retailer, buying agent, vendor, consumer testing service laboratory, and test standard developing agencies associated to the apparel business.

References

1. Das S (2008), 'Salient features of quality evaluation', *Apparel Views*, 7, 65–67.
2. Cullis David (2005), 'Managing apparel warehouses', *Express Textiles*, 10, 9.
3. Das S (2005), 'Value addition to garment', *Apparel Views*, 4, 22–23.
4. AATCC test method135 Dimensional changes of fabrics after home laundering.
5. AATCC test method 150 Dimensional changes of garments after home laundering.
6. AATCC test method 179 Skewness change in fabric and garment twist resulting from automatic home laundering.
7. AATCC test method 61 Colorfastness to laundering: accelerated.
8. AATCC test method 132 Colorfastness to drycleaning.
9. AATCC test method 107 Colorfastness to water.
10. AATCC test method 15 Colorfastness to perspiration.
11. AATCC test method 8 Colorfastness to crocking.
12. AATCC test method 172 Colorfastness to powdered non-chlorine bleach in home laundering.
13. AATCC test method 162 Colorfastness to water: Chlorinated pool.
14. AATCC test method 16 Colorfastness to light.
15. AATCC test method 23 Colorfastness to burnt gas fumes.
16. AATCC test method 109 Colorfastness to ozone in the atmosphere under low humidities.
17. ASTM D 5034 Standard test method for breaking strength and elongation of textile fabrics (Grab Test).
18. ASTM D 1424 Standard test method for tearing strength of fabrics by falling-pendulum (Elmendorf-type) apparatus.
19. ASTM D3786 Standard test method for bursting strength of textile fabrics—Diaphragm bursting strength tester method.
20. ASTM D 3512 Standard test method for pilling resistance and other related surface changes of textile fabrics: Random tumble pilling tester.
21. OEKO-TEX News, edition 01, 2009, Test for harmful substances play significant role in considerations for textile buyers. Available from: www.oeko-tex.com [Accessed 13 February 2009].
22. Doshi Gaurav (2008), Quality control aspects of garment export. Available from: http://Ezine.Articles.com [Accessed 12 February 2009].

CHAPTER 2

Quality protocols and performance standards of apparels and related accessories

Abstract

This chapter discusses performance standards of fabrics and apparels in characterisation of various merchandise products as per their specific application areas and which are acceptable to the retailers in the globe. The chapter first highlights minimum characteristics of varieties of fabrics such as construction, durability, colour fastness, etc., which are essential to its effective use in the garment. The chapter then discusses the performance of a garment in respect of the seams used in different positions of a garment, size and fit properties, appearance, application of accessories, and attachment of various decorations to judge the performance in intended area of application.

Keywords: fabrics, apparels, construction, durability, colour fastness

Chapter contains (Section headings)

2.1 Introduction
2.2 Protocols for apparel testing
    2.2.1 Protocols for zippers, buttons, and snaps testing
    2.2.2 Common testing protocols of woven apparel
    2.2.3 Common testing protocol for knitted apparel
    2.2.4 Common testing protocol for leather/suede apparel
    2.2.5 Common testing protocol for apparel related accessories such as belts, caps, ear muffs, gloves, hats, neckties, scarves, headbands etc.
2.3 Various performance standards of fabrics used in apparel
2.4 Various performance standards in apparel
    2.4.1 Woven shirts, tops, and blouses
    2.4.2 Knit shirts, tops and blouses
    2.4.3 Sweaters

2.1   Introduction

Importance of mechanical and physical properties of fabrics in the clothing manufacturing process[1] has been the subject of many recent investigations in the apparel sector. Testing is essential to characterise the quality of fabrics and apparels.Performance standards cite the test procedures to be used in testing those items. Testing may be done in-house or by an independent third-party laboratory. But one must adhere to the proper use of the test method. The interpretation of the test results will help identify conformance to the standard or otherwise. Fabric performance specifications for various properties and for various end items have been developed. It is true that the buyer and seller must mutually arrive at performance specifications for various properties of an end item, i.e. apparel under consideration. Some reputed retail store chains have their own standard for various clothing items, and also, the test methods to be used are indicated by them. In spite of all the test methods and available test data, the interpretation of test results sometimes governed by a quality decision based on commercial cause.

The protocol in an apparel testing is a summary of requirements of performance, safety, quality and labelled claims. Different parameters are involved in sub-division of different test protocols. Merchandise category and fabric quality govern such type of characterization. While some properties are common for different protocols, additional inputs are essential to properly designate the character of apparel. The matrix of different protocols for apparels and accessories[2] is discussed below.

2.2   Protocols for apparel testing

Label verification:

- Country of origin
- Fibre content

- Care labelling
- Registration (RN) number
- Size
- Copyright verification
- Stuffed articles label (Canada)

Identification tests:

- Fibre analysis
- Yarn size
- Fabric count
- Fabric weight
- Fabric construction

Washability:

- Dimensional stability
- Appearance in laundering (includes: self-staining, torque, skew, trim/ seam durability, trim/garment compatibility, puckering, raspy hand, pill/fuzz, etc.)

Strength and performance tests:

- Tensile (woven)
- Tear (woven)
- Bursting (knit)
- Seam strength/stretchability
- Pocket strength
- Snap/zipper strength
- Stretch and recovery for elastic item
- Pilling
- Pile retention (corduroy)

Colourfastness tests:

- Laundering/drycleaning
- Chlorine bleach
- Non-chlorine bleach
- Crocking
- Light
- Perspiration (lining or skin contact)
- Ozone and burnt gas fume (indigo and white)

Other required tests:

- Flammability
- pH(washed items)
- Azo colourants (European requirement)

Additional test for technical outerwear/rainwear:

- Water repellency
- Water resistance
- Coating verification
- Breathability

Additional test for infant garment:

- Heavy metal/lead content on surface paints/coating
- Formaldehyde content
- Colour fastness to saliva (under 36 months)
- Children safety construction review (includes: small parts, sharp object, drawstring, etc.)

Additional test for intimate and sleepwear:

- Flammability (children sleepwear)
- Yarn slippage
- Colour fastness to perspiration
- Stretch and recovery for elastic band

Additional test for sweaters:

- Garment weight
- Neck stretch

Additional test for swimwear:

- Colour fastness to seawater, water, and chlorinated pool water

Additional test for down fill product:

- Air permeability
- Down proofness
- Down/feather labelling requirement
- Fill power
- Turbidity
- Oxygen number

Additional test for wrinkle resistant garment:

- Formaldehyde content
- Flex abrasion
- Durable press rating

2.2.1   Protocols for zippers, buttons, and snaps testing

General properties:

- Heavy metals (painted surface: Europe and Germany)
- Lead content (all surface coating)
- Nickel leaching (skin contact only)
- Formaldehyde (zipper, button: children under 3 years)

Visual testing:

- Manufacturing qualities (zipper and snap)

Strength/durability properties:

- Appearance after laundering/drycleaning
- Zipper strength
- Button and snap strength
- Resistance to corrosion (metal only)
- Impact resistance (button)
- Centrestrength (button)
- Line size (button)
- Thickness (button)

2.2.2   Common testing protocols of woven apparel

i. Woven natural cellulosic fabrics (50% or more) predominantly made out of cotton, linen, hemp, ramie, jute, and blends,

ii. Woven man-made cellulosic origin (50% or more) predominantly made out of acetate, lyocell, modal, rayon (cuprammonium and viscose), rayon from bamboo (bamboo rayon), triacetate, and blends;

iii. Woven man-made synthetics (50% or more) predominantly made out of acrylic, modacrylic, nylon, aramid, olefin (polyethylene and polypropylene), polyester, spandex, vinal, vinyon, and blends;

iv. Denim; and

v. Woven pile fabrics, i.e. corduroy, terry cloth, velvet, velveteen, and other pile fabrics are given as follows:

- Fabric weight
- Thread count
- Dimensional stability
- Appearance retention
- Skewing
- Tensile strength
- Tearing strength
- Seam strength (production seams)
- Stretch properties (Stretch direction only)
- Colour fastness

2.2.3   Common testing protocol for knitted apparel

- Fabric weight
- Dimensional stability
- Appearance retention
- Skewing
- Bursting strength
- Colour fastness

2.2.4   Common testing protocol for leather/suede apparel

- Dimensional stability
- Appearance retention
- Colour fastness

2.2.5   Common testing protocol for apparel-related accessories such as belts, caps, ear muffs, gloves, hats, neckties, scarves, headbands, etc.

- Fabric weight
- Thickness
- Dimensional stability
- Appearance retention
- Seam strength (production seam)
- Colour fastness

2.3   Various performance standards of fabrics used in apparel

Quality of a fabric plays an important role for specific use in garment[3]. The selection of a fabric to be used in merchandise depends on various physical and chemical parameters. The characteristics of the fabric depend on the type of construction method. Some are more durable than others. The yarns per inch in a fabric are a direct indication of quality. Higher yarn count translates into a higher quality fabric. Higher twist yarns in a fabric are stronger, indicating higher quality fabrics. The fabrics used for interfacings are supportive and build shape and stability in small areas. The fabrics used for underlining add support and durability to the fashion fabric. They are usually decorative and construction details are also important to become compatible with the base fabric. All of these are a significant factor in quality of the garment. Supportive fabrics should be fastened securely, finished appropriately, and should not wrinkle or distort the fabric. Linings should be caught at shoulder seams to prevent slippage and pulling.[3] Linings and fashion fabrics should have compatible care requirements. Fabric testing prevented poor quality garments from being marketed that might otherwise have resulted in damage to the brand image of the companies involved.

Different test methods are there to evaluate the properties of the fabrics used in the garment. It is worthwhile to mention that such properties are determined and compared with the minimum performance standard. The desired standard is benchmarked according to the nature of fabric. Although the performance standard exists for different varieties of fabrics[4], it is normally expected that fabric quality will not only fulfill the standard but also exceed the minimum requirement. Performance standards of different varieties of fabrics are described in Table 2.1 to Table 2.31.

2.4   Various performance standards in apparel

The performance of a garment is regulated by the right quality of fabrics used in the engineering of merchandise. It depends on the seams used in different positions of a garment, size and fit properties, appearance, application of accessories, and attachment of various decorations to judge the performance in actual end use. At each stage of garment engineering, appropriate control measures ensure the production of right quality of apparel for its intended application area. Thus, various apparel performance parameters according to their category of application are described as follows[5].

2.4.1   Woven shirts, tops, and blouses

I.   Fabric

    Fabric constru.ction   As approved/contracted (±5%)
    Flammability   Class 1
    [Must comply with 16 CFR part1610[6] or ASTM D1230[7]]

II.   Garment construction

- Pockets – They will be uniform in size and placed evenly or aligned.
- Seams – They must be finished and back tacked at ends. No untrimmed threads are allowed.They must be free from puckering and correct tension shall be used.
- Stitching – Thread must be colourfast, no broken top stitches, and open seams.
- Darts – They must be uniform in length and shape. There shall be no puckering or bubbles.
- Stress points – They must be bartacked or reinforced as necessary.
- Interfacing – They must have compatible shrinkage to shell fabric and must lie flat.
- Buttons – They must be securely fastened and colourfast. Button holes must be compatible and completely stitched around.
- Snaps, rivets, and trims – They must be securely fastened, reinforced, and no corrosion after 1 hour at rest in laundry machine after one home laundry cycle.
- Elastic and ribbing – They must extend to fullest width of fabric without breaking stitches.Tunnelled elastic must be stitched down to prevent twisting and rollover. No exposed elastic is allowed.
- Stripes and plaids – They must match at all seams unless otherwise specified.
- Plackets – No puckering at seams, especially at bottom or base is desired.
- Hems and edge finishing – They must be even with no raw edges.
- Needle cutting – Correct needle size and type for fabric are required. No needle cuts are expected.
- Zippers – Correct duty zipper for garment is necessary. No bulging or wavering on tape is allowed. Ends of tape must be securely fastened.
- Drawstrings – They must be secured/finished at both ends.
- Children's garments: There must be no hood or neck drawstrings on garments size 2T-12. Waist/Bottom drawstring on age grades 2T-16 may not exceed 3 inches in length outside the drawstring channel when garment is expanded to its fullest width. No toggles, knots, or attachments at the free end are allowed. Drawstrings must be bartacked at centre back so string cannot be pulled out.
- Pile fabrics – There must be no press marks and no crushed pile.
- Painted hardware – There should be less than 0.06% lead by weight[8].

Table 2.1 Minimum performance standards for woven top and mid-weight fabrics – broadcloth, muslin, percale, chambray, poplin, taffeta, and rayon blend

Property	Requirements	Test methods
fibre content		AATCC 20-A
Single fibre	Must be 100% – No foreign fibre	
Multi-fibre	±3.0% of stated fibre content	
Fabric weight	As approval sample ±5%	ASTM D3776
Thread count	As approved sample ±5%	ASTM D3775
Yarn structure	As approved sample	ASTM D1059
Defects	No major defects	ASTM D3990
Flammability		
Adult sleepwear	Class 1	Must comply with 16 CFR-1610
Children's sleepwear	Pass	Must comply with 16 CFR-1615 and 1616
Dimensional stability (3 home launderings)	3% × 3%	AATCC 135 and 150
Tensile (breaking) strength		ASTM D5034/5035
Fabrics (less than 3.5 oz/sq. yd)	20 lbs/in	
Fabrics (3.5 oz./sq. yd or greater)	25 lbs/in	
Tear resistance		ASTM D1424/2261
Fabrics (less than 3.5 oz/sq. yd)	1.5 lbs	
Fabrics (3.5 oz./sq. yd or greater)	3.0 lbs	
Abrasion resistance	50 cycles	ASTM D3886
Appearance ratings		
Pilling resistance	Class 4 @ 30 minutes	ASTM D3514
Smoothness appearance	Class SA 4	AATCC 143

(Contd.)

Property	Requirements	Test methods
Colourfastness ratings		
Colourfastness to laundering-		AATCC 61, 132
Shade change	Class 4.0	
Staining/Bleeding	Class 3.0	
Self staining	Class 4.5	
Chlorine and/or Non-chlorine bleach	Class 4.0	MTL S-1003
Colourfastness to perspiration		AATCC 15
Shade change	Class 4.0	
Staining	Class 3.0	
Colourfastness to light		AATCC 16, option C
Regular fabrics	Class 4.0 min. @ 20 hours	
	Class 4.0 min. @ 10 hours	
100% Nylon – regular colours	Class 4.0 min. @ 20 hours	
	Class 4.0 min. @ 10 hours	
Neon/Fluorescent/ Bright colours	Class 3.0 min. @ 10 hours (Note: Outerwear–Class 2.5–2.0 requires hangtag)	
100% Polyester – regular colours	Class 4.0 min. @ 20 hours	
	Class 4.0 min. @ 10 hours	
Neon/Fluorescent/ Bright colours	Class 3.0 min. @ 20 hours (Note: Outerwear–Class 2.5–2.0 requires hangtag)	
Crocking –		AATCC 8/116
Dry/Wet – original	Class 4.0/3.0	
Dry/Wet – after one wash	Class 4.0/3.0 (must bear advisory hangtag)	

Note: Crocking and bleeding ratings requirements are for light and medium shades. Requirements are reduced one half class for dark shades and pigment prints except when otherwise noted.

Table 2.2   Minimum performance standards for top and mid-weight apparel 100% cotton sheeting

Property	Requirements	Test methods
Fibre content		AATCC 20-A
Single fibre	Must be 100% – no foreign fibre	
Multi-fibre	±3.0% of stated fibre content	
Fabric weight	As approval sample ±5%	ASTM D3776
Thread count	As approved sample ±5%	ASTM D3775
Yarn structure	As approved sample	ASTM D1059
Defects	No major defects	ASTM D3990
Flammability		
Clothing	Class 1	Must comply with 16 CFR-1610
Children's sleepwear	Pass	Must comply with 16 CFR-1615 and 1616
Dimensional stability (3 home launderings)	5% × 5%	AATCC 135 and 150
Tensile (breaking) strength		ASTM D5034/5035
Fabrics (less than 3.5 oz/sq. yd)	20 lbs/in	
Fabrics (3.5 oz./sq. yd or greater)	25 lbs/in	
Tear resistance		ASTM D1424/2261
Fabrics (less than 3.5 oz/sq. yd)	1.5 lbs	
Fabrics (3.5 oz./sq. yd or greater)	3.0 lbs	
Abrasion resistance	50 cycles	ASTM D3886
Seam strength/slippage ¼"		ASTM D1683/434 MOD.
Fabrics (less than 3.5 oz/ sq. yd)	10 lbs	
Fabrics (3.5 oz./sq. yd or greater)	15 lbs	

(Contd.)

Property	Requirements	Test methods
	Appearance ratings	
Pilling resistance	Class 4 @ 30 minutes	ASTM D3514
Smoothness appearance	Class SA 4	AATCC 143
	Colourfastness ratings	
Colourfastness to laundering		AATCC 61, 132
Shade change	Class 3.0	
Staining/Bleeding	Class 3.0	
Self staining	Class 4.5	
Chlorine and/or Non-chlorine bleach	Class 3.0	MTL S-1003
Colourfastness to perspiration		AATCC 15
Shade change	Class 4.0	
Staining	Class 3.0	
Colourfastness to light	Class 4.0 min.@ 20 hours	AATCC 16, option C
	Class 4.0 min.@ 10 hours	
Crocking		AATCC 8/116
Dry/Wet – original	Class 3.5/2.0	
Dry/Wet – after one wash	Class 3.5/2.0 (must bear advisory hangtag)	

Table 2.3   Minimum performance standards for 100% rayon and rayon rich challis, crepe, rib weave, taffeta, and lightweight twill fabrics

Property	Requirements	Test methods
Fibre content		AATCC 20-A
Single fibre	Must be 100% – no foreign fibre	
Multi-fibre	±3.0% of stated fibre content	

(Contd.)

Property	Requirements	Test methods
Fabric weight	As approval sample ±5%	ASTM D3776
Thread count	As approved sample ±5%	ASTM D3775
Yarn structure	As approved sample	ASTM D1059
Defects	No major defects	ASTM D3990
Flammability		
Clothing	Class 1	Must comply with 16 CFR-1610
Children's sleepwear	Pass	Must comply with 16 CFR-1615 and 1616
Dimensional stability		AATCC 135 and 150
(3 home laundering)	5% × 5%	
(commercial drycleaning)	2% × 2%	
Tensile (breaking) strength		ASTM D5034/5035
Fabrics (less than 3.5 oz/sq. yd)	20 lbs/in	
Fabrics (3.5 oz./sq. yd or greater)	25 lbs/in	
Tear resistance		ASTM D1424/2261
Fabrics (less than 3.5 oz/sq. yd)	1.5 lbs	
Fabrics (3.5 oz./sq. yd or greater)	3.0 lbs	
Abrasion resistance	25 cycles	ASTM D3886
	Appearance ratings	
Pilling resistance	Class 4 @ 30 minutes	ASTM D3514
Smoothness appearance	Class SA 4	AATCC 143
	Colourfastness ratings	
Colourfastness to laundering –		AATCC 61, 132
Shade change	Class 4.0	
Staining/Bleeding	Class 3.0	
Self staining	Class 4.5	

(Contd.)

Property	Requirements	Test methods
Chlorine and/or Non-chlorine bleach	Class 4.0	MTL S-1003
Colourfastness to perspiration		AATCC 15
Shade change	Class 4.0	
Staining	Class 3.0	
Colourfastness to light	Class 4.0 min. @ 20 hours	AATCC 16, option C
	Class 4.0 min. @ 10 hours	
Crocking		AATCC 8/116
Dry/Wet – original	Class 4.0/3.0	
Dry/Wet – after one wash	Class 4.0/3.0 (must bear advisory hangtag)	

Note: Crocking and bleeding ratings requirements are for light and medium shades. Requirements are reduced one half class for dark shades and pigment prints except when otherwise noted.

Table 2.4   Minimum performance standards for Indian madras fabrics

Property	Requirements	Test methods
Fibre content		AATCC 20-A
Single fibre	Must be 100% – no foreign fibre	
Multi-fibre	±3.0% of stated fibre content	
Fabric weight	As approval sample ±5%	ASTM D3776
Thread count	As approved sample ±5%	ASTM D3775
Yarn structure	As approved sample	ASTM D1059
Defects	No major defects	ASTM D3990
Flammability		
Clothing	Class 1	Must comply with 16 CFR-1610

(Contd.)

Property	Requirements	Test methods
Children's sleepwear	Pass	Must comply with 16 CFR-1615 and 1616
Dimensional stability (3 home launderings)	8% × 8%	AATCC 135 and 150
Tensile (breaking) strength		ASTM D5034/5035
Fabrics (less than 3.5 oz/sq. yd)	20 lbs/in	
Fabrics (3.5 oz./sq. yd or greater)	25 lbs/in	
Tear resistance		ASTM D1424/2261
Fabrics (less than 3.5 oz/sq. yd)	1.5 lbs	
Fabrics (3.5 oz./sq. yd or greater)	3.0 lbs	
Abrasion resistance		ASTM D3886
Fabrics (less than 3.5 oz/sq. yd)	25 cycles	
Fabrics (3.5 oz./sq. yd or greater)	50 cycles	
Seam strength/ slippage ¼"		ASTM D1683/434 MOD.
Fabrics (less than 3.5 oz/sq. yd)	15 lbs	
Fabrics (3.5 oz./sq. yd or greater)	20 lbs	
Appearance ratings		
Pilling resistance	Class 4 @ 30 minutes	ASTM D3514
Colourfastness ratings		
Colourfastness to laundering		AATCC 61, 132
Shade change	Class 3.0	
Staining/bleeding	Class 2.0	
Self staining	Class 4.5	

(Contd.)

Property	Requirements	Test methods
Chlorine and/or non-chlorine bleach	Class 3.0	MTL S-1003
Colourfastness to perspiration		AATCC 15
Shade change	Class 4.0	
Staining	Class 3.0	
Colourfastness to light	Class 4.0 min. @ 20 hours	AATCC 16, option C
	Class 4.0 min. @ 10 hours	
Crocking		AATCC 8/116
Dry/Wet – original	Class 3.5/2.0	
Dry/Wet – after one wash	Class 3.5/2.0 (must bear advisory hangtag)	

Table 2.5   Minimum performance standards for 100% cotton seersucker

Property	Requirements	Test methods
Fibre content		AATCC 20-A
Single fibre	Must be 100% – no foreign fibre	
Multi-fibre	±3.0% of stated fibre content	
Fabric weight	As approval sample ±5%	ASTM D3776
Thread count	As approved sample ±5%	ASTM D3775
Yarn structure	As approved sample	ASTM D1059
Defects	No major defects	ASTM D3990
Flammability		
Clothing	Class 1	Must comply with 16 CFR-1610
Children's sleepwear	Pass	Must comply with 16 CFR-1615 and 1616

(Contd.)

Property	Requirements	Test methods
Dimensional stability (3 home launderings)	5% × 5%	AATCC 135 and 150
Tensile (breaking) strength		ASTM D5034/5035
Fabrics (less than 3.5 oz/sq. yd)	20 lbs/in	
Fabrics (3.5 oz./sq. yd or greater)	25 lbs/in	
Tear resistance		ASTM D1424/2261
Fabrics (less than 3.5 oz/sq. yd)	1.5 lbs	
Fabrics (3.5 oz./sq. yd or greater)	3.0 lbs	
Abrasion resistance	50 cycles **Appearance ratings**	ASTM D3886
Pilling resistance	Class 4 @ 30 minutes **Colourfastness ratings**	ASTM D3514
Colourfastness to laundering		AATCC 61, 132
Shade change	Class 4.0	
Staining/bleeding	Class 3.0	
Self staining	Class 4.5	
Chlorine and/or non-chlorine bleach	Class 4.0	MTL S-1003
Colourfastness to perspiration		AATCC 15
Shade change	Class 4.0	
Staining	Class 3.0	
Colourfastness to light	Class 4.0 min. @ 20 hours	AATCC 16, option C
	Class 4.0 min. @ 10 hours	
Crocking –		AATCC 8/116

(Contd.)

Property	Requirements	Test methods
Dry/Wet – original	Class 4.0/3.0	
Dry/Wet – after one wash	Class 4.0/3.0 (Must bear advisory hangtag)	

Note: Crocking and bleeding ratings requirements are for light and medium shades. Requirements are reduced one half class for dark shades and pigment prints except when otherwise noted.

Table 2.6   Minimum performance standards for 100% cotton texture weaves i.e. monks cloth, waffle weave, momie, etc

Property	Requirements	Test methods
Fibre content		AATCC 20-A
Single fibre	Must be 100% – no foreign fibre	
Multi fibre	±3.0% of stated fibre content	
Fabric weight	As approval sample ±5%	ASTM D3776
Thread count	As approved sample ±5%	ASTM D3775
Yarn structure	As approved sample	ASTM D1059
Defects	No major defects	ASTM D3990
Flammability		
Clothing	Class 1	Must comply with 16 CFR-1610
Children's sleepwear	Pass	Must comply with 16 CFR-1615 and 1616
Dimensional stability (3 home launderings)		AATCC 135 and 150
Fabrics (less than 3.5 oz/sq. yd)	8% × 8%	
Fabrics (3.5 oz./sq. yd or greater)	6% × 6%	
Tensile (breaking) strength		ASTM D5034/5035
Fabrics (less than 3.5 oz/sq. yd)	25 lbs/in	

(Contd.)

Property	Requirements	Test methods
Fabrics (3.5 oz./sq. yd or greater)	30 lbs/in	
Tear resistance		ASTM D1424/2261
Fabrics (less than 3.5 oz/sq. yd)	1.5 lbs	
Fabrics (3.5 oz./sq. yd or greater)	3.0 lbs	
Abrasion resistance		ASTM D3886
Fabrics (less than 3.5 oz/sq. yd)	25 cycles	
Fabrics (3.5 oz./sq. yd or greater)	50 cycles	
	Appearance ratings	
Pilling resistance	Class 4 @ 30 Minutes	ASTM D3514
	Colourfastness ratings	
Colourfastness to laundering-		AATCC 61, 132
Shade change	Class 4.0	
Staining/bleeding	Class 3.0	
Self staining	Class 4.5	
Chlorine and/or non-chlorine bleach	Class 4.0	MTL S-1003
Colourfastness to perspiration		AATCC 15
Shade change	Class 4.0	
Staining	Class 3.0	
Colourfastness to light	Class 4.0 min. @ 20 hours	AATCC 16, option C
	Class 4.0 min. @ 10 hours	
Crocking-		AATCC 8/116
Dry/Wet – original	Class 4.0/2.0	
Dry/Wet – after one wash	Class 4.0/2.0 (must bear advisory hangtag)	

(Contd.)

Table 2.7   Minimum performance standards for woven gauze fabrics

Property	Requirements	Test methods
Fibre content		AATCC 20-A
Single fibre	Must be 100% – no foreign fibre	
Multi-fibre	±3.0% of stated fibre content	
Fabric weight	As approval sample ±5%	ASTM D3776
Thread count	As approved sample ±5%	ASTM D3775
Yarn structure	As approved sample	ASTM D1059
Defects	No major defects	ASTM D3990
Flammability		
Clothing	Class 1	Must comply with 16 CFR-1610
Children's sleepwear	Pass	Must comply with 16 CFR-1615 and 1616
Dimensional stability (3 home launderings)		AATCC 135 and 150
Fabrics (less than 3.5 oz/sq. yd)	7% × 7%	
Fabrics (3.5 oz./sq. yd or greater)	5% × 5%	
Tensile (breaking) strength		ASTM D5034/5035
Fabrics (less than 3.5 oz/ sq. yd)	20 lbs/in	
Fabrics (3.5 oz./sq. yd or greater)	25 lbs/in	
Tear resistance		ASTM D1424/2261
Fabrics (less than 3.5 oz/sq. yd)	1.5 lbs	
Fabrics (3.5 oz./sq. yd or greater)	3.0 lbs	
Abrasion resistance		ASTM D3886

(Contd.)

Property	Requirements	Test methods
Fabrics (less than 3.5 oz/sq. yd)	25 cycles	
Fabrics (3.5 oz./sq. yd or greater)	50 cycles	
	Appearance ratings	
Pilling resistance	Class 4 @ 30 minutes	ASTM D3514
	Colourfastness ratings	
Colourfastness to laundering		AATCC 61, 132
Shade change	Class 4.0	
Staining/bleeding	Class 3.0	
Self staining	Class 4.5	
chlorine and/or non-chlorine bleach	Class 4.0	MTL S-1003
Colourfastness to perspiration		AATCC 15
Shade change	Class 4.0	
Staining	Class 3.0	
Colourfastness to light	Class 4.0 min. @ 20 hours	AATCC 16, option C
	Class 4.0 min. @ 10 hours	
Crocking		AATCC 8/116
Dry/Wet – original	Class 4.0/2.5	
Dry/Wet – after one wash	Class 4.0/2.5 (must bear advisory hangtag)	

Table 2.8   Minimum performance standards for voile fabrics.

Property	Requirements	Test methods
Fibre content		AATCC 20-A
Single fibre	Must be 100% – no foreign fibre	
Multi fibre	±3.0% of stated fibre content	

(Contd.)

Property	Requirements	Test methods
Fabric weight	As approval sample ±5%	ASTM D3776
Thread count	As approved sample ±5%	ASTM D3775
Yarn structure	As approved sample	ASTM D1059
Defects	No major defects	ASTM D3990
Flammability		
Clothing	Class 1	Must comply with 16 CFR-1610
Children's sleepwear	Pass	Must comply with 16 CFR-1615 and 1616
Dimensional stability (3 home launderings)		AATCC 135 and 150
Cotton/Synthetic blends	3% × 3%	
100 % Cotton fabrics	5% × 5% (must fit labeled size range)	
Tensile (breaking) strength		ASTM D5034/5035
Fabrics (less than 3.5 oz/sq. yd)	20 lbs/in	
Fabrics (3.5 oz./sq. yd or greater)	25 lbs/in	
Tear resistance		ASTM D1424/2261
Fabrics (less than 3.5 oz/sq. yd)	1.5 lbs	
Fabrics (3.5 oz./sq. yd or greater)	3.0 lbs	
Abrasion resistance		ASTM D3886
Fabrics (less than 3.5 oz/sq. yd)	25 cycles	
Fabrics (3.5 oz./sq. yd or greater)	50 cycles	
	Appearance ratings	
Pilling resistance	Class 4 @ 30 Minutes	ASTM D3514
	Colourfastness ratings	
Colourfastness to laundering		AATCC 61, 132

(Contd.)

Property	Requirements	Test methods
Shade change	Class 4.0	
Staining/bleeding	Class 3.0	
Self staining	Class 4.5	
Chlorine and/or non-chlorine bleach	Class 4.0	MTL S-1003
Colourfastness to perspiration		AATCC 15
Shade change	Class 4.0	
Staining	Class 3.0	
Colourfastness to light	Class 4.0 min. @ 20 hours	AATCC 16, option C
	Class 4.0 min. @ 10 hours	
Crocking		AATCC 8/116
Dry/Wet – original	Class 4.0/3.0	
Dry/Wet – after one wash	Class 4.0/3.0 (must bear advisory hangtag)	

Note: Crocking and bleeding ratings requirements are for light and medium shades. Requirements are reduced one half class for dark shades and pigment prints except when otherwise noted.

Table 2.9 Minimum performance standards for pigment printed flannel shirting fabrics.

Property	Requirements	Test methods
Fibre content		AATCC 20-A
Single fibre	Must be 100% – no foreign fibre	
Multi-fibre	±3.0% of stated fibre content	
Fabric weight	As approval sample ±5%	ASTM D3776
Thread count	As approved sample ±5%	ASTM D3775
Yarn structure	As approved sample	ASTM D1059
Defects	No major defects	ASTM D3990

(Contd.)

Property	Requirements	Test methods
Flammability		
Clothing	Class 1	Must comply with 16 CFR-1610
Children's sleepwear	Pass	Must comply with 16 CFR-1615 and 1616
Dimensional stability (3 home launderings)	5% × 5%	AATCC 135 and 150
Tensile (breaking) strength	25 lbs/in	ASTM D5034/5035
Tear resistance	3.0 lbs	ASTM D1424/2261
Abrasion resistance	50 cycles	ASTM D3886
Appearance ratings		
Pilling resistance	Class 3 @ 30 minutes	ASTM D3514
Colourfastness ratings		
Colourfastness to laundering		**AATCC 61, 132**
Shade change	Class 3.0	
Staining/bleeding	Class 3.0	
Self staining	Class 4.5	
Chlorine and/or Non-chlorine bleach	Class 3.5	MTL S-1003
Colourfastness to perspiration		AATCC 15
Shade change	Class 3.0	
Staining	Class 3.0	
Colourfastness to light	Class 4.0 min. @ 20 hours	AATCC 16, option C
	Class 4.0 min. @ 10 hours	
Crocking		AATCC 8/116
Dry/Wet – original	Class 3.0/2.0	
Dry/Wet – after one wash	Class 3.0/2.0 (must bear advisory hangtag)	

(Contd.)

Table 2.10 Minimum performance standards for yarn dyed flannel shirting fabrics.

Property	Requirements	Test methods
Fibre content		AATCC 20-A
Single fibre	Must be 100% - no foreign fibre	
Multi-fibre	±3.0% of stated fibre content	
Fabric weight	As approval sample ±5%	ASTM D3776
Thread count	As approved sample ±5%	ASTM D3775
Yarn structure	As approved sample	ASTM D1059
Defects	No major defects	ASTM D3990
Flammability		
Clothing	Class 1	Must comply with 16 CFR-1610
Children's sleepwear	Pass	Must comply with 16 CFR-1615 and 1616
Dimensional stability (3 home launderings)	5% × 5%	AATCC 135 and 150
Tensile (breaking) strength	25 lbs/in	ASTM D5034/5035
Tear resistance	3.0 lbs	ASTM D1424/2261
Abrasion resistance	50 cycles	ASTM D3886
	Appearance ratings	
Pilling resistance	Class 3 @ 30 minutes	ASTM D3514
	Colourfastness ratings	
Colourfastness to laundering		AATCC 61, 132
Shade change	Class 4.0	
Staining/bleeding	Class 3.0	
Self staining	Class 4.5	
Chlorine and/or non-chlorine bleach	Class 3.5	MTL S-1003
Colourfastness to perspiration		AATCC 15

(Contd.)

Property	Requirements	Test methods
Shade change	Class 3.0	
Staining	Class 3.0	
Colourfastness to light	Class 4.0 min. @ 20 hours	AATCC 16, option C
	Class 4.0 min. @ 10 hours	
Crocking		AATCC 8/116
Dry/Wet – original	Class 3.0/2.5	
Dry/Wet – after one wash	Class 3.0/2.5 (must bear advisory hangtag)	

Table 2.11   Minimum performance standards for stretch twills and denim.

Property	Requirements	Test methods
Fibre content		AATCC 20-A
Single fibre	Must be 100% – no foreign fibre	
Multi-fibre	±3.0% of stated fibre content	
Fabric weight	As approval sample ±5%	ASTM D3776
Thread count	As approved sample ±5%	ASTM D3775
Yarn structure	As approved sample	ASTM D1059
Defects	No major defects	ASTM D3990
Flammability		
Clothing	Class 1	Must comply with 16 CFR-1610
Children's sleepwear	Pass	Must comply with 16 CFR-1615 and 1616
Dimensional stability (3 home launderings)	3% × 3%	AATCC 135 and 150
Torque/twisting	5% (Based on length)	AATCC 179
Tensile (breaking) strength	25 lbs/in	ASTM D5034/5035
Tear resistance	3.0 lbs	ASTM D1424/2261
Abrasion resistance	100 cycles	ASTM D3886
pH balance (Garment wash programs)	pH between 6 and 8	AATCC 81

(Contd.)

Property	Requirements	Test methods
Seam strength/Slippage ¼"	25 lbs/in	ASTM D1683/434 MOD.
	Appearance ratings	
Pilling resistance	Class 3 @ 30 minutes	ASTM D3514
Smoothness appearance	Class SA 3.5	AATCC 143
	Colourfastness ratings	
Colourfastness to laundering		AATCC 61, 132
Shade alteration	Class 4.0	
Staining	Class 4.0	
Bleeding	Class 3.0	
Self staining	Class 4.5	
Chlorine &/or Non-chlorine bleach	Class 4.0	MTL S-1003
Colourfastness to perspiration		AATCC 15
Shade change	Class 4.0	
Staining	Class 3.0	
Colourfastness to light	Class 4.0 min. @ 20 hours	AATCC 16, option C
	Class 4.0 min. @ 10 hours	
Crocking		AATCC 8/116
Dry/Wet – original	Class 4.0/2.0	
Dry/Wet – after one wash	Class 4.0/2.0 (must bear advisory hangtag)	

Table 2.12 Minimum performance standards for bottom weight twills, duck and canvas ($\geq$ 8 oz/sq yd).

Property	Requirements	Test methods
Fibre content		AATCC 20-A
Single fibre	Must be 100% – no foreign fibre	
Multi-fibre	±3.0% of stated fibre content	
Fabric weight	As approval sample ±5%	ASTM D3776

(Contd.)

Property	Requirements	Test methods
Thread count	As approved sample ±5%	ASTM D3775
Yarn structure	As approved sample	ASTM D1059
Defects	No major defects	ASTM D3990
Flammability		
Clothing	Class 1	Must comply with 16 CFR-1610
Children's sleepwear	Pass	Must comply with 16 CFR-1615 and 1616
Dimensional stability (3 home launderings)	3% × 3%	AATCC 135 and 150
Torque/twisting	5% (based on length)	AATCC 179
Tensile (breaking) strength	25 lbs/in	ASTM D5034/5035
Tear resistance	3.0 lbs	ASTM D1424/2261
Abrasion resistance	Unwashed 500 cycles	ASTM D3886
	Garment washed 200 cycles	
pH balance (Garment wash programs)	pH between 6 and 8	AATCC 81
Seam strength/Slippage ¼"	30 lbs/in	ASTM D1683/434 MOD.
	Appearance ratings	
Pilling resistance	Class 4 @ 30 minutes	ASTM D3514
Smoothness appearance	Class SA 4	AATCC 143
	Colourfastness ratings	
Colourfastness to Laundering-		AATCC 61, 132
Shade change	Class 4.0	
Staining	Class 4.0	
Bleeding	Class 3.0	
Self staining	Class 4.5	
Chlorine and/or non-chlorine bleach	Class 4.0	MTL S-1003
Colourfastness to perspiration		AATCC 15
Shade change	Class 4.0	

(Contd.)

Property	Requirements	Test methods
Staining	Class 3.0	
Colourfastness to light	Class 4.0 min. @ 20 hours	AATCC 16, option C
	Class 4.0 min. @ 10 hours	
Crocking		AATCC 8/116
Dry/Wet – original	Class 4.0/2.0	
Dry/Wet – after one wash	Class 4.0/2.0 (must bear advisory hangtag)	

Table 2.13 Minimum performance standards for bottom weight denim (≥ 8 oz/sq yd).

Property	Requirements	Test methods
Fibre content		AATCC 20-A
Single fibre	Must be 100% – no foreign fibre	
Multi-fibre	±3.0% of stated fibre content	
Fabric weight	As approval sample ±5%	ASTM D3776
Thread count	As approved sample ±5%	ASTM D3775
Yarn structure	As approved sample	ASTM D1059
Defects	No major defects	ASTM D3990
Flammability		
Clothing	Class 1	Must comply with 16 CFR-1610
Children's sleepwear	Pass	Must comply with 16 CFR-1615 and 1616
Dimensional stability (3 home launderings)	3% × 3%	AATCC 135 and 150
Torque/Twisting	5% (based on length)	AATCC 179
Tensile (breaking) strength	50 lbs/in	ASTM D5034/5035
Tear resistance	3.0 lbs	ASTM D1424/2261
Abrasion resistance	Unwashed 500 cycles	ASTM D3886

(Contd.)

Property	Requirements	Test methods
	Garment washed 200 cycles	
pH balance (Garment wash programs)	pH between 6 and 8	AATCC 81
Seam strength/Slippage ¼"	Unwashed 50 lbs/in	ASTM D1683/434 MOD.
	Garment washed 40 lbs/in	
	Appearance ratings	
Pilling resistance	Class 4 @ 30 minutes	ASTM D3514
Smoothness appearance	Class SA 4	AATCC 143
	Colourfastness ratings	
Colourfastness to laundering		AATCC 61, 132
Shade change	Class 4.0	
Staining	Class 3.0	
Bleeding	Class 3.0	
Self staining	Class 4.5	
Chlorine and/or non-chlorine bleach	Class 4.0	MTL S-1003
Colourfastness to perspiration		AATCC 15
Shade change	Class 4.0	
Staining	Class 3.0	
Colourfastness to Light	Class 4.0 min. @ 20 hours	AATCC 16, option C
	Class 4.0 min. @ 10 hours	
Crocking		AATCC 8/116
Dry/Wet – original	Class 4.0/2.0	
Dry/Wet – after one wash	Class 4.0/2.0 (must bear advisory hangtag)	

Note: Minimum performances standards apply to finished fabric/ garment; i.e., after stone/acid/enzyme wash, etc.

Table 2.14 Minimum performance standards for corduroy, velveteen, and velvet fabrics (includes flocked velvet).

Property	Requirements	Test methods
Fibre content		AATCC 20-A
Single fibre	Must be 100% – no foreign fibre	
Multi-fibre	±3.0% of stated fibre content	
Fabric weight	As approval sample ±5%	ASTM D3776
Thread count	As approved sample ±5%	ASTM D3775
Yarn structure	As approved sample	ASTM D1059
Defects	No major defects	ASTM D3990
Flammability		
Clothing	Class 1	Must comply with 16 CFR-1610
Children's sleepwear	Pass	Must comply with 16 CFR-1615 and 1616
Dimensional stability		AATCC 135 and 150
(3 home launderings)	5% × 5%	
(commercial drycleaning)	2% × 2%	
Tensile (breaking) strength	25 lbs/in	ASTM D5034/5035
Tear resistance	3.0 lbs	ASTM D1424/2261
	Appearance Ratings	
Pile retention (cut pile fabrics only)		ASTM D4685
Fabrics (less than 7.0 oz/sq. yd)	Face : Class 3 @ 300 cycles	
	Back : Class 3 @ 50 cycles	
Fabrics (7.0 oz./sq. yd or greater)	Face : Class 3 @ 300 cycles	

(Contd.)

Property	Requirements	Test methods
After wash appearance (After drycleaning or 3 home launderings)	Back : Class 3 @ 100 cycles	AATCC 135
Pile appearance/ retention	No appreciable picking or bare spots	
	Colourfastness ratings	
Colourfastness to laundering		AATCC 61, 132
Shade change	Class 3.5	
Staining/bleeding	Class 3.0	
Self staining	Class 4.5	
Chlorine and/or Non-chlorine bleach	Class 3.5	MTL S-1003
Colourfastness to perspiration		AATCC 15
Shade change	Class 3.0	
Staining	Class 3.0	
Colourfastness to light	Class 4.0 min. @ 20 hours Class 4.0 min. @ 10 hours	AATCC 16, option C
Crocking		AATCC 8/116
Dry/Wet – original	Class 3.0/2.0	
Dry/Wet – after one wash	Class 3.0/2.0 (must bear advisory hangtag)	

Table 2.15   Minimum performance standards for woven terry cloth and chenille fabrics for garment.

Property	Requirements	Test methods
Fibre content		AATCC 20-A
Single fibre	Must be 100% – no foreign fibre	
Multi-fibre	±3.0% of stated fibre content	
Fabric weight	As approval sample ±5%	ASTM D3776

(Contd.)

Property	Requirements	Test methods
Thread count	As approved sample ±5%	ASTM D3775
Yarn structure	As approved sample	ASTM D1059
Defects	No major defects	ASTM D3990
Flammability		
Clothing	Class 1	Must comply with 16 CFR-1610
Children's sleepwear	Pass	Must comply with 16 CFR-1615 and 1616
Dimensional stability (5 and 10 home launderings)	7% × 5%	AATCC 135 and 150
Tensile (breaking) strength	25 lbs/in	ASTM D5034/5035
Tear resistance	3.0 lbs	ASTM D1424/2261
	Appearance ratings	
Pile retention (cut pile fabrics only)		ASTM D4685
Fabrics (less than 7.0 oz/sq. yd)	Face : Class 3 @ 300 cycles	
	Back : Class 3 @ 50 cycles	
Fabrics (7.0 oz./sq. yd or greater)	Face : Class 3 @ 300 cycles	
	Back : Class 3 @ 100 cycles	
After wash appearance (After 10 home launderings)		AATCC 135
Seam durability	No raveling	
Pile Appearance/ retention	No appreciable picking or bare spots	
	Colourfastness ratings	
Colourfastness to laundering		AATCC 61, 132
Shade change	Class 3.5	
Staining/bleeding	Class 3.0	
Self staining	Class 4.5	

(Contd.)

Property	Requirements	Test methods
Chlorine and/or non-chlorine bleach	Class 3.5	MTL S-1003
Colourfastness to perspiration		AATCC 15
Shade change	Class 3.0	
Staining	Class 3.0	
Colourfastness to Light	Class 4.0 min. @ 20 hours	AATCC 16, option C
	Class 4.0 min. @ 10 hours	
Crocking		AATCC 8/116
Dry/Wet – original	Class 3.0/2.0	
Dry/Wet – after one wash	Class 3.0/2.0 (must bear advisory hangtag)	

Table 2.16 Minimum performance standards for wool blend.

Property	Requirements	Test methods
Fibre content		AATCC 20-A
Single fibre	Must be 100% – no foreign fibre	
Multi-fibre	±3.0% of stated fibre content	
Fabric weight	As approval sample ±5%	ASTM D3776
Thread count	As approved sample ±5%	ASTM D3775/3887
Yarn structure	As approved sample	ASTM D1059
Defects	No major defects	ASTM D3990
Flammability		
Clothing	Class 1	Must comply with 16 CFR-1610
Children's sleepwear	Pass	Must comply with 16 CFR-1615 and 1616
Dimensional stability (3 home launderings)		AATCC 135 and 150

(Contd.)

Property	Requirements	Test methods
Woven fabrics	3% X 3%	
Knit fabrics	5% X 5%	
Commercial drycleaning	2% X 2%	
Tensile (breaking) strength		ASTM D5034/5035
Fabrics (less than 3.5 oz/sq. yd)	20 lbs/in	
Fabrics (3.5 oz./sq. yd or greater)	25 lbs/in	
Tear resistance		ASTM D1424/2261
Fabrics (less than 3.5 oz/sq. yd)	1.5 lbs	
Fabrics (3.5 oz./sq. yd or greater)	3.0 lbs	
Bursting strength (Knit fabrics)		ASTM D3786
Fabrics (less than 3.5 oz/sq. yd)	40 psi	
Fabrics (3.5 oz./sq. yd or greater)	55 psi	
Abrasion resistance	50 cycles	ASTM D3886
Seam strength/Slippage ¼"		ASTM D1683/434 MOD.
Fabrics (less than 3.5 oz/sq. yd)	15 lbs	
Fabrics (3.5 oz./sq. yd or greater)	20 lbs	
	Appearance ratings	
Pilling resistance	Class 3 @ 30 minutes	ASTM D3512/3514
Smoothness appearance	Class SA 4	AATCC 143
	Colourfastness ratings	
Colourfastness to laundering		AATCC 61, 132
Shade change	Class 4.0	
Staining/Bleeding	Class 3.0	

(Contd.)

Property	Requirements	Test methods
Self staining	Class 4.5	
Chlorine and/or non-chlorine bleach	Class 4.0	MTL S-1003
Colourfastness to perspiration		AATCC 15
Shade change	Class 4.0	
Staining	Class 3.0	
Colourfastness to light	Class 4.0 min. @ 20 hours	AATCC 16, option C
	Class 4.0 min. @ 10 hours	
Crocking		AATCC 8/116
Dry/Wet – original	Class 4.0/3.0	
Dry/Wet – after one wash	Class 4.0/3.0 (must bear advisory hangtag)	

Note: Crocking and bleeding ratings requirements are for light and medium shades. Requirements are reduced one half class for dark shades and pigment prints except when otherwise noted.

Table 2.17   Minimum performance standards for lining fabrics.

Property	Requirements	Test methods
Fibre content		AATCC 20-A
Single fibre	Must be 100% – no foreign fibre	
Multi-fibre	±3.0% of stated fibre content	
Fabric weight	As approval sample ±5%	ASTM D3776
Thread count	As approved sample ±5%	ASTM D3775/3887
Yarn structure	As approved sample	ASTM D1059
Defects	No major defects	ASTM D3990
Flammability		
Clothing	Class 1	Must comply with 16 CFR-1610

(Contd.)

Property	Requirements	Test methods
Children's sleepwear	Pass	Must comply with 16 CFR-1615 and 1616
Dimensional stability (3 home launderings)		AATCC 135 and 150
Woven fabrics	3% × 3%	
Knit fabrics	5% × 5%	
Commercial drycleaning	2% X 2%	
Differential shrinkage (between shell and lining)	2%	
Tensile (breaking) strength		ASTM D5034/5035
Fabrics (less than 3.5 oz/sq. yd)	20 lbs/in	
Fabrics (3.5 oz./sq. yd or greater)	25 lbs/in	
Tear resistance		ASTM D1424/2261
Fabrics (less than 3.5 oz/sq. yd)	1.5 lbs	
Fabrics (3.5 oz./sq. yd or greater)	3.0 lbs	
Bursting strength (knit fabrics)		ASTM D3786
Fabrics (less than 3.5 oz/sq. yd)	40 psi	
Fabrics (3.5 oz./sq. yd or greater)	55 psi	
Abrasion resistance	25 cycles	ASTM D3886
Seam strength/slippage ¼"		ASTM D1683/434 MOD.
Fabrics (less than 3.5 oz/sq. yd)	15 lbs	
Fabrics (3.5 oz./sq. yd or greater)	20 lbs	
	Appearance ratings	
Pilling resistance	Class 4 @ 30 minutes	ASTM D3512/3514

(Contd.)

Property	Requirements	Test methods
	Colourfastness ratings	
Colourfastness to laundering		AATCC 61, 132
Shade change	Class 4.0	
Staining/Bleeding	Class 3.0	
Self staining	Class 4.5	
Chlorine and/or Non-chlorine bleach	Class 4.0	MTL S-1003
Colourfastness to perspiration		AATCC 15
Shade change	Class 4.0	
Staining	Class 3.0	
Colourfastness to light	Class 4.0 min. @ 20 hours	AATCC 16, option C
	Class 4.0 min. @ 10 hours	
Crocking		AATCC 8/116
Dry/Wet – original	Class 4.0/3.0	
Dry/Wet – after one wash	Class 4.0/3.0 (must bear advisory hangtag)	

Note: Crocking and bleeding ratings requirements are for light and medium shades. Requirements are reduced one half class for dark shades and pigment prints except when otherwise noted.

Table 2.18   Minimum performance standards for woven silk fabrics.

Property	Requirements	Test methods
Fibre content		AATCC 20-A
Single fibre	Must be 100% – No foreign fibre	
Multi fibre	±3.0% of stated fibre content	
Fabric weight	As approval sample ±5%	ASTM D3776
Thread count	As approved sample ±5%	ASTM D3775

(Contd.)

Property	Requirements	Test methods
Yarn structure	As approved sample	ASTM D1059
Defects	No major defects	ASTM D3990
Slubs, holes, misweaves, etc. may not be discernible from one foot way.		
Fabric weight definition	Ounces/square yard	Momme
Heavy	> 4.1	> 31.5
Medium	2.6 – 4.0	20.5 – 31.4
Light	1.6 – 2.5	12.6 – 20.4
Sheer	≤ 1.5	≤ 12.5
Flammability		
Clothing	Class 1	Must comply with 16 CFR-1610
Children's sleepwear	Pass	Must comply with 16 CFR-1615 and 1616
Dimensional stability (3 home launderings)	5% × 5%	AATCC 135 and 150
Commercial drycleaning	2% × 2%	
Tensile (breaking) strength		ASTM D5034/5035
- Heavy/Med.	20 lbs/in	
- Light	15 lbs/in	
- Sheer	10 lbs/in	
Tear resistance		ASTM D1424/2261
- Heavy/Med.	2.0 lbs	
- Light	1.5 lbs	
- Sheer	1.0 lbs	
Abrasion resistance		ASTM D3886
- Heavy/Med.	50 cycles	
- Light	25 cycles	
- Sheer	N/A	
Seam strength/slippage ¼"		ASTM D1683/434 MOD.
- Heavy/Med.	15 lbs	

(Contd.)

Property	Requirements	Test methods
- Light	10 lbs	
- Sheer	10 lbs	
	Appearance ratings	
After wash appearance	Good after wash appearance	AATCC 135
(After drycleaning or 3 launderings)		
	Colourfastness ratings	
Colourfastness to laundering		AATCC 61, 132
Shade change	Class 4.0	
Staining/bleeding	Class 2.5	
Self staining	Class 4.5	
Chlorine and/or non-chlorine bleach	Class 4.0	MTL S-1003
Colourfastness to perspiration		AATCC 15
Shade change	Class 4.0	
Staining	Class 3.0	
Colourfastness to light	Class 4.0 min. @ 10 Hours	AATCC 16, Option C
Crocking -		AATCC 8/116
Dry/Wet – original	Class 4.0/2.0	
Dry/Wet – after one wash	Class 4.0/2.0 (must bear advisory hangtag)	

Table 2.19   Minimum performance standards for knit silk fabrics.

Property	Requirements	Test methods
Fibre content		AATCC 20-A
Single fibre	Must be 100% – no foreign fibre	
Multi-fibre	±3.0% of stated fibre content	
Fabric weight	As approval sample ±5%	ASTM D3776

(Contd.)

Property	Requirements	Test methods
Thread count	As approved sample ±5%	ASTM D3887
Yarn structure	As approved sample	ASTM D1059
Defects	No major defects	ASTM D3990
Slubs, holes, misweaves, etc. may not be discernible from one feet way.		
Fabric weight definition	Ounces/square yard	Momme
Heavy	> 4.1	> 31.5
Medium	2.6 – 4.0	20.5 – 31.5
Light	1.6 – 2.5	12.6 – 20.4
Sheer	≤ 1.5	≤ 12.5
Flammability		
Clothing	Class 1	Must comply with 16 CFR-1610
Children's sleepwear	Pass	Must comply with 16 CFR-1615 and 1616
Dimensional stability (3 home launderings)	5% × 5%	AATCC 135 and 150
Commercial drycleaning	2% × 2%	
Bursting strength		ASTM D3786
- Heavy/med.	50 psi	
- Light	40 psi	
- Sheer	30 psi	
Abrasion resistance		ASTM D3886
- Heavy/med.	50 cycles	
- Light	25 cycles	
- Sheer	N/A	
Seam elongation		ASTM D1683
- Heavy/med.	50% elongation or 7.0 lbs tension	
- Light	30% elongation or 6.0 lbs tension	
	Appearance ratings	
After wash appearance	Good after wash appearance	AATCC 135

(Contd.)

Property	Requirements	Test methods
(After drycleaning or 3 launderings)		
	Colourfastness ratings	
Colourfastness to laundering		AATCC 61, 132
Shade change	Class 4.0	
Staining/bleeding	Class 2.5	
Self staining	Class 4.5	
Chlorine and/or non-chlorine bleach	Class 4.0	MTL S-1003
Colourfastness to perspiration		AATCC 15
Shade change	Class 4.0	
Staining	Class 3.0	
Colourfastness to light	Class 4.0 min. @ 10 hours	AATCC 16, Option C
Crocking		AATCC 8/116
Dry/Wet – original	Class 4.0/2.0	
Dry/Wet – after one wash	Class 4.0/2.0 (must bear advisory hangtag)	

Table 2.20 Minimum performance standards for chief value cellulose (CVC = 51% or greater cellulose) jersey and interlock fabrics.

Property	Requirements	Test methods
Fibre content		AATCC 20-A
Single fibre	Must be 100% – no foreign fibre	
Multi-fibre	±3.0% of stated fibre content	
Fabric weight	As approval sample ±5%	ASTM D3776
Thread count	As approved sample ±5%	ASTM D3887
Yarn structure	As approved sample	ASTM D1059
Defects	No major defects	ASTM D3990

(Contd.)

Property	Requirements	Test methods
Flammability		
Clothing	Class 1	Must comply with 16 CFR-1610
Children's sleepwear	Pass	Must comply with 16 CFR-1615 and 1616
Dimensional stability (3 home launderings)	7% × 7%	AATCC 135/150
Torque/Twisting	5% of length	AATCC 179
Bursting strength		ASTM D3786
Fabrics (less than 3.5 oz/sq. yd)	40 psi	
Fabrics (3.5 oz./sq. yd or greater)	55 psi	
Abrasion resistance		ASTM D3886
Fabrics (less than 3.5 oz/sq. yd)	25 cycles	
Fabrics (3.5 oz./sq. yd or greater)	50 cycles	
	Appearance ratings	
Pilling resistance	Class 3 @ 30 minutes	ASTM D3512
Smoothness appearance	Class SA 4	AATCC 143
	Colourfastness ratings	
Colourfastness to laundering		AATCC 61, 132
Shade change	Class 4.0	
Staining/bleeding	Class 3.0	
Self staining	Class 4.5	
Chlorine and/or Non-chlorine bleach	Class 4.0	MTL S-1003
Colourfastness to perspiration		AATCC 15
Shade change	Class 4.0	
Staining	Class 3.0	
Colourfastness to light	Class 4.0 min. @ 20 hours	AATCC 16, Option C

(Contd.)

Property	Requirements	Test methods
	Class 4.0 min. @ 10 hours	
Crocking		AATCC 8/116
Dry/Wet – original	Class 4.0/3.0	
Dry/Wet – after one wash	Class 4.0/3.0 (must bear advisory hangtag)	

Note: Crocking and bleeding ratings requirements are for light and medium shades. Requirements are reduced one half class for dark shades and pigment prints except when otherwise noted.

Table 2.21 Minimum performance standards for chief value synthetic (CVS = 51% or greater synthetic) jersey and interlock knit fabrics.

Property	Requirements	Test methods
Fibre content		AATCC 20-A
Single fibre	Must be 100% – No foreign fibre	
Multi-fibre	±3.0% of stated fibre content	
Fabric weight	As approval sample ±5%	ASTM D3776
Thread count	As approved sample ±5%	ASTM D3887
Yarn structure	As approved sample	ASTM D1059
Defects	No major defects	ASTM D3990
Flammability		
Clothing	Class 1	Must comply with 16 CFR-1610
Children's sleepwear	Pass	Must comply with 16 CFR-1615 and 1616
Dimensional stability (3 home launderings)	5% × 5%	AATCC 135/150
Torque/twisting	5% of length	AATCC 179
Bursting strength		ASTM D3786
Fabrics (less than 3.5 oz/sq. yd)	40 psi	

(Contd.)

Property	Requirements	Test methods
Fabrics (3.5 oz./sq. yd or greater)	55 psi	
Abrasion resistance		ASTM D3886
Fabrics (less than 3.5 oz/sq. yd)	25 cycles	
Fabrics (3.5 oz./sq. yd or greater)	50 cycles	
	Appearance ratings	
Pilling resistance	Class 3 @ 30 minutes	ASTM D3512
Smoothness appearance	Class SA 4	AATCC 143
	Colourfastness ratings	
Colourfastness to laundering		AATCC 61, 132
Shade change	Class 4.0	
Staining/bleeding	Class 3.0	
Self staining	Class 4.5	
Chlorine and/or non-chlorine bleach	Class 4.0	MTL S-1003
Colourfastness to perspiration		AATCC 15
Shade change	Class 4.0	
Staining	Class 3.0	
Colourfastness to light	Class 4.0 min. @ 20 hours	AATCC 16, Option C
	Class 4.0 min. @ 10 hours	
Crocking		AATCC 8/116
Dry/Wet – original	Class 4.0/3.0	
Dry/Wet – after one wash	Class 4.0/3.0 (must bear advisory hangtag)	

Note: Crocking and bleeding ratings requirements are for light and medium shades. Requirements are reduced one half class for dark shades and pigment prints except when otherwise noted.

Table 2.22 Minimum performance standards for chief value cellulose (CVC = 51% or greater cellulose) rib knit fabrics

Property	Requirements	Test methods
Fibre content		AATCC 20-A
Single fibre	Must be 100% – no foreign fibre	
Multi-fibre	±3.0% of stated fibre content	
Fabric weight	As approval sample ±5%	ASTM D3776
Thread count	As approved sample ±5%	ASTM D3887
Yarn structure	As approved sample	ASTM D1059
Defects	No major defects	ASTM D3990
Flammability		
Clothing	Class 1	Must comply with 16 CFR-1610
Children's sleepwear	Pass	Must comply with 16 CFR-1615 and 1616
Dimensional stability (3 home launderings)	7% × 10%	AATCC 135/150
Torque/Twisting	5% of length	AATCC 179
Bursting strength		ASTM D3786
Fabrics (less than 3.5 oz/sq. yd)	40 psi	
Fabrics (3.5 oz./sq. yd or greater)	55 psi	
Abrasion resistance		ASTM D3886
Fabrics (less than 3.5 oz/sq. yd)	25 cycles	
Fabrics (3.5 oz./sq. yd or greater)	50 cycles	
	Appearance ratings	
Pilling resistance	Class 3 @ 30 minutes	ASTM D3512
Smoothness appearance	Class SA 4	AATCC 143
	Colourfastness ratings	
Colourfastness to laundering		AATCC 61, 132

(Contd.)

Property	Requirements	Test methods
Shade change	Class 4.0	
Staining/bleeding	Class 3.0	
Self staining	Class 4.5	
Chlorine and/or Non-chlorine bleach	Class 4.0	MTL S-1003
Colourfastness to perspiration		AATCC 15
Shade change	Class 4.0	
Staining	Class 3.0	
Colourfastness to light	Class 4.0 min. @ 20 hours	AATCC 16, Option C
	Class 4.0 min. @ 10 hours	
Crocking		AATCC 8/116
Dry/Wet – original	Class 4.0/3.0	
Dry/Wet – after one wash	Class 4.0/3.0 (must bear advisory hangtag)	

Note: Crocking and bleeding ratings requirements are for light and medium shades. Requirements are reduced one half class for dark shades and pigment prints except when otherwise noted.

Table 2.23 Minimum performance standards for chief value synthetic (CVS = 51% or greater synthetic) rib knit fabrics

Property	Requirements	Test methods
Fibre content		AATCC 20-A
Single fibre	Must be 100% – no foreign fibre	
Multi-fibre	±3.0% of stated fibre content	
Fabric weight	As approval sample ± 5%	ASTM D3776
Thread count	As approved sample ±5%	ASTM D3887
Yarn structure	As approved sample	ASTM D1059
Defects	No major defects	ASTM D3990
Flammability		
Clothing	Class 1	Must comply with 16 CFR-1610

(Contd.)

Property	Requirements	Test methods
Children's sleepwear	Pass	Must comply with 16 CFR-1615 and 1616
Dimensional stability (3 home launderings)	5% ×8%	AATCC 135
Torque/twisting	5% of length	AATCC 179
Bursting strength		ASTM D3786
Fabrics (less than 3.5 oz/sq. yd)	40 psi	
Fabrics (3.5 oz./sq. yd or greater)	55 psi	
Abrasion resistance		ASTM D3886
Fabrics (less than 3.5 oz/sq. yd)	25 cycles	
Fabrics (3.5 oz./sq. yd or greater)	50 cycles	
	Appearance ratings	
Pilling resistance	Class 3 @ 30 minutes	ASTM D3512
Smoothness appearance	Class SA 4	AATCC 143
	Colourfastness ratings	
Colourfastness to laundering		AATCC 61, 132
Shade change	Class 4.0	
Staining/bleeding	Class 3.0	
Self staining	Class 4.5	
Chlorine and/or non-chlorine bleach	Class 4.0	MTL S-1003
Colourfastness to perspiration-		AATCC 15
Shade change	Class 4.0	
Staining	Class 3.0	
Colourfastness to light	Class 4.0 min. @ 20 hours	AATCC 16, Option C
	Class 4.0 min. @ 10 hours	
Crocking		AATCC 8/116
Dry/Wet – original	Class 4.0/3.0	

(Contd.)

Property	Requirements	Test methods
Dry/Wet – after one wash	Class 4.0/3.0 (must bear advisory hangtag)	

Note: Crocking and bleeding ratings requirements are for light and medium shades. Requirements are reduced one half class for dark shades and pigment prints except when otherwise noted.

Table 2.24 Minimum performance standards for stretch knit fabrics with spandex

Property	Requirements	Test methods
Fibre content		AATCC 20-A
Single fibre	Must be 100% – no foreign fibre	
Multi-fibre	±3.0% of stated fibre content	
Fabric weight	As approval sample ±5%	ASTM D3776
Thread count	As approved sample ±5%	ASTM D3887
Yarn structure	As approved sample	ASTM D1059
Defects	No major defects	ASTM D3990
Flammability		
Clothing	Class 1	Must comply with 16 CFR-1610
Children's sleepwear	Pass	Must comply with 16 CFR-1615 and 1616
Dimensional stability (3 home launderings)		AATCC 135
CVC	8% × 8%	
CVC rib	8% × 11%	
CVS	6% × 6%	
CVS rib	6% × 9%	
Torque/Twisting	5% of length	AATCC 179
Bursting strength		ASTM D3786
Fabrics (less than 3.5 oz/sq. yd)	40 psi	
Fabrics (3.5 oz./sq. yd or greater)	55 psi	

(Contd.)

Property	Requirements	Test methods
Abrasion resistance		ASTM D3886
Fabrics (less than 3.5 oz/sq. yd)	25 cycles	
Fabrics (3.5 oz/sq. yd or greater)	50 cycles	
	Appearance ratings	
Pilling resistance	Class 3 @ 30 minutes	ASTM D3512
Smoothness appearance	Class SA 4	AATCC 143
	Colourfastness ratings	
Colourfastness to laundering		AATCC 61, 132
Shade change	Class 4.0	
Staining/bleeding	Class 3.0	
Self staining	Class 4.5	
Chlorine and/or non-chlorine bleach	Class 4.0	MTL S-1003
Colourfastness to perspiration –		AATCC 15
Shade change/Staining	Class 4.0/3.0	
Colourfastness to sea water –		AATCC 106
Shade change/staining	Class 4.0/3.0	
Colourfastness to chlorinate pool water		AATCC 162
Shade change	Class 4.0	
Colourfastness to light		AATCC 16, Option C
Regular colours	Class 4.0 min. @ 20 hours	
	Class 4.0 min. @ 10 hours	
Neon/fluorescent/bright colours	Class 3.0 min. @ 10 hours (Note: Class 2.5 – 2.0 requires hangtag)	
Crocking		AATCC 8/116
Dry/Wet – original	Class 4.0/3.0	
Dry/Wet – after one wash	Class 4.0/3.0(must bear advisory hangtag)	

Note: Crocking and bleeding ratings requirements are for light and medium shades. Requirements are reduced one half class for dark shades and pigment prints except when otherwise noted.

Table 2.25 Minimum performance standards for CVC knit fleece and French terry fabrics

Property	Requirements	Test methods
Fibre content		AATCC 20-A
Single fibre	Must be 100% – no foreign fibre	
Multi-fibre	±3.0% of stated fibre content	
Fabric weight	As approval sample ±5%	ASTM D3776
Thread count	As approved sample ±5%	ASTM D3887
Yarn structure	As approved sample	ASTM D1059
Defects	No major defects	ASTM D3990
Flammability		
Clothing	Class 1	Must comply with 16 CFR-1610
Children's sleepwear	Pass	Must comply with 16 CFR-1615 and 1616
Dimensional stability (3 home launderings)	8% × 8%	AATCC 135 and 150
Torque/twisting	5% of length	AATCC 179
Bursting strength	55 psi	ASTM D3786
Abrasion resistance	50 cycles	ASTM D3886
	Appearance ratings	
Pilling resistance	Class 3 @ 30 minutes	ASTM D3512
Smoothness appearance	Class SA 4	AATCC 143
	Colourfastness ratings	
Colourfastness to laundering		AATCC 61, 132
Shade change	Class 4.0	
Staining/bleeding	Class 3.0	
Self staining	Class 4.5	
Chlorine and/or non-chlorine bleach	Class 4.0	MTL S-1003
Colourfastness to perspiration –		AATCC 15
Shade change	Class 4.0	

(Contd.)

Property	Requirements	Test methods
Staining	Class 3.0	
Colourfastness to light	Class 4.0 min. @ 20 hours Class 4.0 min. @ 10 hours	AATCC 16, Option C
Crocking		AATCC 8/116
Dry/Wet – original	Class 3.5/3.0	
Dry/Wet – after one wash	Class 3.5/3.0 (must bear advisory hangtag)	

Note: Crocking and bleeding ratings requirements are for light and medium shades. Requirements are reduced one half class for dark shades and pigment prints except when otherwise noted.

Table 2.26   Minimum performance standards for CVS knit fleece, French terry and polar type fleece fabrics.

Property	Requirements	Test methods
Fibre content		AATCC 20-A
Single fibre	Must be 100% – no foreign fibre	
Multi-fibre	±3.0% of stated fibre content	
Fabric weight	As approval sample ±5%	ASTM D3776
Thread count	As approved sample ±5%	ASTM D3887
Yarn structure	As approved sample	ASTM D1059
Defects	No major defects	ASTM D3990
Flammability		
Clothing	Class 1	Must comply with 16 CFR-1610
Children's sleepwear	Pass	Must comply with 16 CFR-1615 and 1616
Dimensional stability (3 home launderings)	5% × 5%	AATCC 135 and 150
Torque/Twisting	5% of length	AATCC 179
Bursting strength	55 psi Appearance ratings	ASTM D3786

(Contd.)

Property	Requirements	Test methods
Pilling resistance	Class 3 @ 30 minutes (Good after wash appearance)	ASTM D3512
	Colourfastness ratings	
Colourfastness to laundering		AATCC 61, 132
Shade change	Class 4.0	
Staining/bleeding	Class 3.0	
Self staining	Class 4.5	
Chlorine and/or non-chlorine bleach	Class 4.0	MTL S-1003
Colourfastness to perspiration		AATCC 15
Shade change	Class 4.0	
Staining	Class 3.0	
Colourfastness to light	Class 4.0 min. @ 20 hours Class 4.0 min. @ 10 Hours	AATCC 16, Option C
Crocking		AATCC 8/116
Dry/Wet – original	Class 4.0/3.0	
Dry/Wet – after one wash	Class 4.0/3.0 (must bear advisory hangtag)	

Note: Crocking and bleeding ratings requirements are for light and medium shades. Requirements are reduced one half class for dark shades and pigment prints except when otherwise noted.

Table 2.27 Minimum performance standards for thermal knit fabrics

Property	Requirements	Test methods
Fibre content		AATCC 20-A
Single fibre	Must be 100% – no foreign fibre	
Multi-fibre	±3.0% of stated fibre content	
Fabric weight	As approval sample ±5%	ASTM D3776
Thread count	As approved sample ±5%	ASTM D3887
Yarn structure	As approved sample	ASTM D1059
Defects	No major defects	ASTM D3990
Flammability		

(Contd.)

Property	Requirements	Test methods
Clothing	Class 1	Must comply with 16 CFR-1610
Children's sleepwear	Pass	Must comply with 16 CFR-1615 and 1616
Dimensional stability (3 home launderings – must restore to fit)		AATCC 135 and 150
Chief value cotton	10% × 10%	
Chief value synthetic	7% × 7%	
Torque/twisting	5% of length	AATCC 179
Bursting strength	55 psi	ASTM D3786
Abrasion resistance	50 cycles	ASTM D3886
	Appearance ratings	
Pilling resistance	Class 3 @ 30 minutes	ASTM D3512
Smoothness appearance	Class SA 4	AATCC 143
	Colourfastness ratings	
Colourfastness to laundering		AATCC 61, 132
Shade change	Class 4.0	
Staining/bleeding	Class 3.0	
Self staining	Class 4.5	
Chlorine and/or non-chlorine bleach	Class 4.0	MTL S-1003
Colourfastness to perspiration		AATCC 15
Shade change	Class 4.0	
Staining	Class 3.0	
Colourfastness to light	Class 4.0 min. @ 20 hours	AATCC 16, Option C
	Class 4.0 min. @ 10 hours	
Crocking		AATCC 8/116
Dry/Wet – original	Class 4.0/3.0	
Dry/Wet – after one wash	Class 4.0/3.0 (must bear advisory hangtag)	

Note: Crocking and bleeding ratings requirements are for light and medium shades. Requirements are reduced one half class for dark shades and pigment prints except when otherwise noted.

Table 2.28 Minimum performance standards for texture knit novelties (pebble, popcorn, waffle and other texture knits formed using dropped stitches to create open effects)

Property	Requirements	Test methods
Fibre content		AATCC 20-A
Single fibre	Must be 100% – No foreign fibre	
Multi-fibre	±3.0% of stated fibre content	
Fabric weight	As approval sample ±5%	ASTM D3776
Thread count	As approved sample ±5%	ASTM D3887
Yarn structure	As approved sample	ASTM D1059
Defects	No major defects	ASTM D3990
Flammability		
Clothing	Class 1	Must comply with 16 CFR-1610
Children's sleepwear	Pass	Must comply with 16 CFR-1615 and 1616
Dimensional stability (3 home launderings – must restore to fit)		AATCC 135 and 150
CVC	8% × 8%	
CVC rib	8% × 11%	
CVS	6% × 6%	
CVS rib	6% × 9%	
Torque/twisting	5% of length	AATCC 179
Bursting strength	55 psi	ASTM D3786
Abrasion resistance	50 cycles	ASTM D3886
	Appearance ratings	
Pilling resistance	Class 3 @ 30 minutes	ASTM D3512
Smoothness appearance	Class SA 4	AATCC 143
	Colourfastness ratings	
Colourfastness to laundering		AATCC 61, 132
Shade change	Class 4.0	
Staining/bleeding	Class 3.0	

(Contd.)

Self staining	Class 4.5	
Chlorine and/or non-chlorine bleach	Class 4.0	MTL S-1003
Colourfastness to perspiration		AATCC 15
Shade change	Class 4.0	
Staining	Class 3.0	
Colourfastness to light	Class 4.0 min. @ 20 hours Class 4.0 min. @ 10 hours	AATCC 16, Option C
Crocking		AATCC 8/116
Dry/Wet – original	Class 4.0/3.0	
Dry/Wet – after one wash	Class 4.0/3.0 (must bear advisory hangtag)	

Note: Crocking and bleeding ratings requirements are for light and medium shades. Requirements are reduced one half class for dark shades and pigment prints except when otherwise noted.

Table 2.29　Minimum performance standards for chief value synthetic (CVS) warp knit fabrics.

Property	Requirements	Test methods
Fibre content		AATCC 20-A
Single fibre	Must be 100% – no foreign fibre	
Multi-fibre	±3.0% of stated fibre content	
Fabric weight	As approval sample ±5%	ASTM D3776
Thread count	As approved sample ±5%	ASTM D3887
Yarn structure	As approved sample	ASTM D1059
Defects	No major defects	ASTM D3990
Flammability		
Clothing	Class 1	Must comply with 16 CFR-1610
Children's sleepwear	Pass	Must comply with 16 CFR-1615 and 1616
Dimensional stability (3 home launderings)	5% × 5%	AATCC 135 and 150

(Contd.)

Property	Requirements	Test methods
Torque/twisting	5% of length	AATCC 179
Bursting strength		ASTM D3786
Fabrics (less than 3.5 oz/sq. yd)	40 psi	
Fabrics (3.5 oz./sq. yd or greater)	55 psi	
Abrasion resistance		ASTM D3886
Fabrics (less than 3.5 oz/sq. yd)	25 cycles	
Fabrics (3.5 oz./sq. yd or greater)	50 cycles	
	Appearance ratings	
Pilling resistance	Class 4 @ 30 minutes	ASTM D3512
Smoothness appearance	Class SA 4	AATCC 143
	Colourfastness ratings	
Colourfastness to laundering		AATCC 61, 132
Shade change	Class 4.0	
Staining/bleeding	Class 3.0	
Self staining	Class 4.5	
Chlorine and/or non-chlorine bleach	Class 4.0	MTL S-1003
Colourfastness to perspiration		AATCC 15
Shade change	Class 4.0	
Staining	Class 3.0	
Colourfastness to light	Class 4.0 min. @ 20 hours	AATCC 16, Option C
	Class 4.0 min. @ 10 hours	
Crocking		AATCC 8/116
Dry/Wet – original	Class 4.0/3.0	
Dry/Wet – after one wash	Class 4.0/3.0 (must bear advisory hangtag)	

Note: Crocking and bleeding ratings requirements are for light and medium shades. Requirements are reduced one half class for dark shades and pigment prints except when otherwise noted.

Table 2.30   Minimum performance standards for knit CVS swimwear fabrics

Property	Requirements	Test methods
Fibre content		AATCC 20-A
Single fibre	Must be 100% – no foreign fibre	
Multi fibre	±3.0% of stated fibre content	
Fabric weight	As approval sample ±5%	ASTM D3776
Thread count	As approved sample ±5%	ASTM D3887
Yarn structure	As approved sample	ASTM D1059
Defects	No major defects	ASTM D3990
Flammability	Class 1	Must comply with 16 CFR-1610
Dimensional stability (3 home launderings – must restore to fit)	5% × 5%	AATCC 135 and 150
Torque/twisting	5% of length	AATCC 179
Bursting strength	55 psi	ASTM D3786
Abrasion resistance	50 cycles	ASTM D3886
	Appearance ratings	
Pilling resistance	Class 4 @ 30 minutes	ASTM D3512
Elasticity/recovery	No change following laundry, perspiration or static water testing	
	Colourfastness ratings	
Colourfastness to laundering		AATCC 61, 132
Shade change	Class 4.0	
Staining/bleeding	Class 3.0	
Self staining	Class 4.5	
Chlorine and/or non-chlorine bleach	Class 4.0 (when appropriate)	MTL S-1003
Colourfastness to perspiration		AATCC 15
Shade change/staining	Class 4.0/4.0	
Colourfastness to sea water –		AATCC 106
Shade change/Staining	Class 4.0/4.0	

(Contd.)

Colourfastness to chlorinate pool water –		AATCC 162
Shade change	Class 4.0	
Colourfastness to light		AATCC 16, Option C
Regular colours	Class 4.0 min. @ 20 hours	
	Class 4.0 min. @ 10 hours	
Neon/fluorescent/bright colours	Class 3.0 min. @ 10 hours (Note: Class 2.5 – 2.0 requires hangtag)	
Crocking		AATCC 8/116
Dry/Wet – original	Class 4.0/3.0	
Dry/Wet – after one wash	Class 4.0/3.0 (must bear advisory hangtag)	

Note: Crocking and bleeding ratings requirements are for light and medium shades. Requirements are reduced one half class for dark shades and pigment prints except when otherwise noted.

Table 2.31   Minimum performance standards for woven swimwear fabrics.

Property	Requirements	Test methods
Fibre content		AATCC 20-A
Single fibre	Must be 100% – no foreign fibre	
Multi-fibre	±3.0% of stated fibre content	
Fabric weight	As approval sample ±5%	ASTM D3776
Thread count	As approved sample ±5%	ASTM D3775
Yarn structure	As approved sample	ASTM D1059
Defects	No major defects	ASTM D3990
Flammability	Class 1	Must comply with 16 CFR-1610
Dimensional stability (3 home launderings)	3% × 3%	AATCC 135 and 150
Tensile (breaking) strength	25 lbs/in	ASTM D5034/5035
Tear resistance	3.0 lbs	ASTM D1424/2261
Abrasion resistance	50 cycles	ASTM D3886
	Appearance ratings	
Pilling resistance	Class 4 @ 30 minutes	ASTM D3514

(Contd.)

Elasticity/recovery	No Change following laundry, pool water and sea water testing	
	Colourfastness ratings	
Colourfastness to laundering		AATCC 61, 132
Shade change	Class 4.0	
Staining/Bleeding	Class 3.0	
Self staining	Class 4.5	
Chlorine and/or non-chlorine bleach	Class 4.0 (when appropriate)	MTL S-1003
Colourfastness to perspiration		AATCC 15
Shade change/staining	Class 4.0/4.0	
Colourfastness to sea water		AATCC 106
Shade change/staining	Class 4.0/4.0	
Colourfastness to chlorinate pool water		AATCC 162
Shade change	Class 4.0	
Colourfastness to light		AATCC 16, Option C
Regular colours	Class 4.0 min. @ 20 hours	
	Class 4.0 min. @ 10 hours	
Neon/fluorescent/bright colours	Class 3.0 min. @ 10 hours (Note; Class 2.5 – 2.0 requires hangtag)	
Crocking-		AATCC 8/116
Dry/Wet – original	Class 4.0/3.0	
Dry/Wet – after one wash	Class 4.0/3.0 (must bear advisory hangtag)	

Note: Crocking and bleeding ratings requirements are for light and medium shades. Requirements are reduced one half class for dark shades and pigment prints except when otherwise noted.

III. Care labelling

These items must meet all requirements of the care labelling rule. Provision of practical care instructions is required.

- All care labels must remain legible and securely attached through the laundering cycle and for the life of the garment.

- Labelling will be determined by the testing lab upon completion of all colourways of actual production sample.
- All trims and hardware must be able to withstand selected care method.
- Labels must be bi-lingual (English/Spanish).

IV. Product labelling

The garment must be labelled with:

- Brand name
- Fibre content
- Garment size
- Country of origin
- RN# or WPL#; or the name and address of the manufacturer or distributor.

V. Size and fit

- Technical Designer shall verify and approve as per desired specification.

VI. Performance standards

A. Fabric performance properties – Appropriate fabric performance standards as mentioned in section 2.3.

B. Seam performance properties

Test	Method	Requirement
Woven seams	ASTM D 1683[9]	
Fabric < 3.5 oz/sq yd		15 lbs/in
Fabric ≥ 3.5 oz/sq yd		20 lbs/in

C. Strength at stress points and of applied decorations

Test	Method	Requirement
Reinforced stress points	ASTM D 1683 (Modified)	20 lbs
Applied decorations	AST ASTM D 1683 (Modified)	
(i) for 18-month age and up		15 lbs min @ 10 sec
(ii) for under 18-month age		10 lbs. min @ 10 sec
Pocket strength	ASTM D 1683 (Modified) Non-functional	
Under 8 oz/sq yd	Functional	10 lbs
	Non-functional	15 lbs
Over 8 oz/sq yd	Functional	10 lbs
		20 lbs

D. Button and snap strength properties

Test	Method	Requirement
Anchored strength	ASTM D 1683 (Modified)	
(i) for 18-month age and up		15 lbs min @10 sec
(ii) for under 18-month age		10 lbs min @10 sec
Button impact resistance	MTL S-1001[10]	5.5 in oz (no failure noted)
Snap/desnap	ASTM D 4846[11]	Opening 2–5 lbs Closing 2–10 lbs

E. Zipper strength properties

Test	Method	Requirement
Seams	ASTM D 2061 (Modified)[12]	20 lbs
Top/bottom stop (open and closed position)		15 lbs

F. Appearance

Appearance tests include the change in garment appearance after repeated washings. This includes but is not limited to:

- Corrosion resistance (metal hardware)   Must display no corrosion
- Lead content of painted hardware   < 0.06%
- Ironing (If recommended)   No colour change
  Good appearance
  Good shape retention
  - Good dimensional stability as per appropriate fabric standard.
  - Retention of fabric smoothness and recovery properties.
  - No chipping, discolouration, or rusting of hardware.
  - No fraying or ravelling of buttonholes, trims, or hems.
  - No twisting of seams.
  - No seam openings or needle cuts.
  - No objectionable frosting, pilling, and snagging.
  - No deterioration of elasticised areas.
  - No deterioration of buttons, snaps, or zippers.

- Torque shall be less than 5% of side seam length.
- Differential shrinkage shall be no more than 2% between shell and lining

G. Smoothness appearance

Test	Method	Requirement
Fabric	AATCC 143[13]	Class 4
Seam appearance	AATCC 143	Class 3

2.4.2 Knit shirts, tops, and blouses

I. Fabric

Fabric construction   As approved/contracted ($\pm$5%)
Flammability   Class 1
(Must comply with 16 CFR part1610 or ASTM D1230)

II. Garment construction

- Pockets – They will be uniform in size and placed evenly or aligned.
- Seams – They must be finished and back tacked at ends. No untrimmed threads are allowed. They must be free from puckering appropriate tension, and seam type shall be used. They must stretch with fabric without breaking.
- Shoulder seams – They must be taped or reinforced unless otherwise specified.
- Stitching – Thread must be colourfast. No broken top stitches, open seams, and breakage are allowed when fabric is fully extended.
- Darts – They must be uniform in length and shape and no puckering or bubbles.
- Stress points – They must be bartacked or reinforced as necessary.
- Interfacing – They must have compatible shrinkage to shell fabric and must lie flat.
- Buttons – They must be securely fastened and colourfast. Button holes must be compatible and completely stitched around.
- Snaps, rivets, and trims – They must be securely fastened, reinforced, and no corrosion after 1 hour at rest in laundry machine after one home laundry cycle.
- Elastic and ribbing – They must extend to fullest width of fabric without breaking stitches. Tunnelled elastic must be stitched down to prevent twisting and rollover. No exposed elastic is allowed.

- Stripes and plaids – They must match at all seams unless otherwise specified.
- Plackets – No puckering at seams is allowed, especially at bottom or base.
- Hems and edge finishing – They must be even with no raw edges.
- Needle cutting– Correct needle size and type for fabric are necessary. Ball point needles should be used with knits to prevent needle cutting.
- Zippers – Correct duty zipper for garment is necessary.No bulging or wavering on tape is allowed.Ends of tape must be securely fastened.
- Drawstrings – They must be secured/finished at both ends. Children's garments: There must be no hood or neck drawstrings on garments size 2T-12. Waist/Bottom drawstring on age grades 2T-16 may not exceed 3 inches in length outside the drawstring channel when garment is expanded to its fullest width.No toggles, knots, or attachments at the free end are allowed. Drawstrings must be bartacked at centre back so string cannot be pulled out.
- Pile fabrics – There must be no press marks and no crushed pile.
- Painted hardware – There should be less than 0.06% lead by weight

III. Care labelling

These items must meet all requirements of the Care labelling rule. Provision of practical care instructions is required.

- All care labels must remain legible and securely attached through the laundering cycle and for the life of the garment.
- They will be determined by the testing lab upon completion of all colourways of actual production sample.
- All trims and hardware must be able to withstand selected care method.
- Labels must be bi-lingual (English/Spanish).

IV. Product labelling: The garment must be labelled with

- Brand name
- Fibrecontent
- Garment size
- Country of origin
- RN# or WPL#; or the name and address of the manufacturer or distributor.

V. Size and fit

- Technical Designer shall verify and approve as per desired specification.

VI. *Performance standards*

A. Fabric performance properties– Appropriate fabric performance standards as mentioned in section 2.3.

B. Seam performance properties

Test	Method	Requirement
Knit and non-woven seams	ASTM D 1683 (Modified)	
Fabric < 3.5 oz/sq yd		30% Elongationor 6 lbs tension
Fabric≥ 3.5 oz/sq yd		50% Elongationor 7 lbs tension

C. Strength at stress points and of applied decorations

Test	Method	Requirement
Reinforced stress points	ASTM D 1683 (Modified)	20 lbs
Applied decorations	AST ASTM D 1683 (Modified)	
(i) for 18-month age and up		15 lbs min @ 10 sec
(ii) for under 18-month age		10 lbs. min @ 10 sec
Pocket strength	ASTM D 1683 (Modified)	
	Non-functional	5 lbs
	Functional	10 lbs

D. Snap and button strength properties

Test	Method	Requirement
Anchored strength	ASTM D 1683 (Modified)	
(i) for 18-month age and up and		15 lbs min @10 sec
(ii) for under 18-month age		10 lbs min @10 sec
Button impact resistance	MTL S-1001	5.5 in oz (no failure noted)
Snap/desnap	ASTM D 4846	Opening 2–5 lbs Closing 2–10 lbs

E. Zipper strength properties

Test	Method	Requirement
Seams	ASTM D 2061 (Modified)	20 lbs
Top/bottom stop(open and closed position)		15 lbs

F. Appearance

Appearance tests include the change in garment appearance after repeated washings. This includes but is not limited to:

- Corrosion resistance (metal hardware).   Must display no corrosion
- Lead content of painted hardware   < 0.06%
- Ironing (If recommended)   No colour change
                                             Good appearance
                                             Good shape retention.

  - Good dimensional stability as per appropriate fabric standard.
  - Retention of fabric stretch and recovery properties.
  - No chipping, discolouration or rusting of hardware.
  - No fraying or ravelling of buttonholes, trims, or hems.
  - No twisting of seams.
  - No seam openings.
  - No objectionable frosting, pilling, and snagging.
  - No deterioration of elasticised areas.
  - No deterioration of buttons, snaps, or zippers.
  - Torque shall be less than 5% of side seam length.
  - Differential shrinkage shall be no more than 2% between shell and lining.

2.4.3 Sweaters

I. Fabric

Fabric construction   As approved/contracted ($\pm$5%)
Flammability   Class 1
(Must comply with 16 CFR part1610 or ASTM D1230 required)

II. Garment construction

- Seams – They must be finished and back tacked at ends. No untrimmed threads are allowed. They must be free from puckering appropriate tension and seam type shall be used. They must stretch with fabric without breaking
- Stretch and recovery–Fabric must stretch as per specification and recover to original shape.
- Shoulder seams– Stretch knits must be taped or reinforced unless otherwise specified.
- Stitching– Thread must be colourfast with no broken top stitches, open seams, and no breakage when fabric is fully extended.

- Stress points – They must be reinforced or bartacked as necessary.
- Buttons– They must be securely fastened and colourfast. Button holes must be compatible and completely stitched around.
- Snaps, rivets, and trims– They must be securely fastened, reinforced, and no corrosion after 1 hour at rest in laundry machine after one home laundry cycle.
- Elastic and ribbing– They must extend to fullest width of fabric without breaking stitches. Tunnelled elastic must be stitched down to prevent twisting and rollover. No exposed elastic is allowed. They must be restored to body size after drycleaning or home laundering.
- Stripes, patterns, and plaids – They must match at all seams unless otherwise specified.
- Plackets– No puckering at seams is allowed, especially at bottom or base.
- Drawstrings – They must be secured/finished at both ends. Children's garments: There must be no hood or neck drawstrings on garments size 2T-12. Waist/Bottom drawstring on age grades 2T-16 may not exceed 3 inches in length outside the drawstring channel when garment is expanded to its fullest width. No toggles, knots, or attachments at the free end are allowed. Drawstrings must be bartacked at centre back so string cannot be pulled out.
- Hems and edge finishing – They must be even with no raw edges.
- Pockets – They must be uniform in size and placed evenly or aligned.
- Needle cutting– Correct needle size and type for fabric are required. Ball point needles should be used with knits to prevent needle cutting.
- General appearance– No visible snags, runs, or loose yarns and holes at colour changes/joining are allowed.
- Zippers – Correct duty zipper for garment is necessary. No bulging or wavering on tape is allowed. Ends of tape must be securely fastened.
- Pile fabrics – There must be no press marks and crushed pile.
- Painted hardware– There should be less than 0.06% lead by weight

III. *Care labelling*

These items must meet all requirements of the care labelling rule. Provision of practical care instructions is required.

- All care labels must remain legible and securely attached through the laundering cycle and for the life of the garment.

- They will be determined by the testing lab upon completion of all colourways of actual production sample.
- All trims and hardware must be able to withstand selected care method.
- Labels must be bi-lingual (English/Spanish).

IV. *Product labelling:The garment must be labelled with*

- Brand name
- Fibre content
- Garment size
- Country of origin
- RN# or WPL#; or the name and address of the manufacturer or distributor.
- Labels must be bi-lingual (English/Spanish).

V. *Size and fit*

- Technical Designer shall verify and approve as per requirement.

VI. *Performance standards*

A. Fabric performance properties– **A**ppropriate fabric performance standards as mentioned in section 2.3.

B. Seam performance properties

Test	Method	Requirement
Knit and non-woven seams	ASTM D 1683 (Modified)	
Fabric <3.5 oz/sq yd		30% Elongationor 6 lbs tension
Fabric ≥3.5 oz/sq yd		50% Elongationor 7 lbs tension

C. Strength at stress points and of applied decorations

Test	Method	Requirement
Reinforced stress points	ASTM D 1683 (Modified)	20 lbs
Applied decorations	ASTASTM D 1683	
(i) for 18-month age and up	(Modified)	15 lbs min @ 10 sec
(ii) for under 18-month age		10 lbs. min @ 10 sec
Pocket strength	ASTM D 1683 (Modified)	
	Non-functional	5 lbs
	Functional	10 lbs

D. Snap and button strength properties

Test	Method	Requirement
Anchored strength (i) for 18-month age and up and (ii) for under 18-month age	ASTM D 1683 (Modified)	15 lbs min @10 sec 10 lbs min @10 sec
Button impact resistance	MTL S-1001	5.5 in oz (no failure noted)
Snap/desnap	ASTM D 4846	Opening 2–5 lbs Closing 2–10 lbs

E. Zipper strength properties

Test	Method	Requirement
Seams	ASTM D 2061 (Modified)	20 lbs
Top/bottom stop(open and closed position)	ASTM D 2061	15 lbs

F. Appearance

Appearance tests include the change in garment appearance after repeated washings. This includes but is not limited to:

- Corrosion resistance (metal hardware)   Must display no corrosion
- Lead content of painted hardware   < 0.06%
- Ironing (If recommended)   No colour change
  Good appearance
  Good shape retention.

  – Good dimensional stability (See appropriate fabric standards).
  – Retention of fabric stretch and recovery properties.
  – No chipping, discolouration or rusting of hardware.
  – No fraying or ravelling of buttonholes, trims, or hems.
  – No twisting of seams.
  – No seam openings.
  – No objectionable frosting, pilling, and snagging.
  – No deterioration of elasticised areas.
  – No deterioration of buttons, snaps, or zippers.
  – No frayed ends shall be visible. Knots/joining must remain secure.
  – Torque must be less than 5% of side seam length.
  – Differential shrinkage of more than 2% between shell and lining is not allowed.

2.4.4 Woven slacks, pants, and shorts

I. Fabric

Fabric construction   As approved/contracted ($\pm5\%$)
Flammability   Class 1 required
(Must comply with 16 CFR part1610 or ASTM D1230)

II. Garment construction

- Pockets – They will be uniform size and placed evenly or aligned.
- Seams – They must be finished and back tacked at ends. No untrimmed threads are allowed. They must be free from puckering, and correct tension shall be used.
- Stitching – Thread must be colourfast, no broken top stitches, and open seams.
- Darts – They must be uniform in length and shape. There shall be no puckering or bubbles.
- Stress points – They must be bartacked or reinforced as necessary.
- Buttons – They must be securely fastened and colourfast. Button holes must be compatible and completely stitched around.
- Snaps, rivets, and trims – They must be securely fastened, reinforced, and no corrosion after 1 hour at rest in laundry machine after one home laundry cycle.
- Elastic and ribbing – They must extend to fullest width of fabric without breaking stitches. Tunnelled elastic must be stitched down to prevent twisting and rollover. No exposed elastic is allowed.
- Stripes and plaids – They must match at all seams unless otherwise specified.
- Hems and edge finishing – There must be even with no raw or unfinished edges.
- Needle cutting – Correct needle size and type for fabric are required. No needle cuts are expected.
- Zippers – Correct duty zipper for garment is necessary. No bulging or wavering on tape is allowed. Ends of tape,slider, and teeth must be securely fastened.
- Drawstrings – They must be secured/finished at both ends.
- Children's garments:There must be no hood or neck drawstrings on garments size 2T-12. Waist/Bottom drawstring on age grades 2T-16 may not exceed 3 inches in length outside the drawstring channel when garment is expanded to its fullest width. No toggles, knots, or attachments at the free end are allowed. Drawstrings must be bartacked at centre back so string cannot be pulled out.
- Pile fabrics – No press marks. No crushed pile.
- Painted hardware – Must be less than 0.06% lead by weight.

III. Care labelling

These items must meet all requirements of the care labelling rule. Provision of practical care instructions is required.

- All care labels must remain legible and securely attached through the laundering cycle and for the life of the garment.
- They will be determined by the testing lab upon completion of all colourways of actual production sample.
- All trims and hardware must be able to withstand selected care method.
- Labels must be bi-lingual (English/Spanish).

IV. Product labelling: The garment must be labelled with

- Brand name
- Fibre content
- Garment size
- Country of origin
- RN# or WPL#; or the name and address of the manufacturer. or distributor.

VI. Size and fit

- Technical Designer shall verify and approve as per requirement.

VI. Performance standards

A. Fabric performance properties – Appropriate fabric performance standards as mentioned in section 2.3.

B. Seam performance properties

Test	Method	Requirement
Woven seams	ASTM D 1683	
(i) Fabric <3.5 oz/sq yd		15 lbs/in
(ii) Fabric ≥3.5 oz/sq yd		20 lbs/in
(iii) 10 oz or greater denimsand twills (felled or double stitched seams)		50 lbs/in

C. Strength at stress points and of applied decorations

Test	Method	Requirement
Reinforced stress points	ASTM D 1683 (Modified)	20 lbs
Applied decorations	ASTM D 1683 (Modified)	
(i) for 18-month age and up		15 lbs min @ 10 sec
(ii) for under 18-month age		10 lbs. min @ 10 sec

(Contd.)

Test	Method	Requirement
Pocket strength	ASTM D 1683 (Modified)	
(i) Under 8 oz/sq yd	Non-functional	10 lbs
	Functional	15 lbs
(ii) Over 8 oz/sq yd	Non-functional	10 lbs
	Functional	20 lbs

D. Snap and button strength properties

Test	Method	Requirement
Anchored strength	ASTM D 1683 (Modified)	
(i) for 18-month age and up and		15 lbs min @10 sec
(ii) for under 18-month age		10 lbs min @10 sec
Button impact resistance	MTL S-1001	5.5 in oz (no failure noted)
Snap/desnap	ASTM D 4846	Opening 2–5 lbs Closing2–10 lbs

E. Zipper strength properties

Test	Method	Requirement
Seams	ASTM D 2061 (Modified)	20 lbs
Top/bottom stop(open and closed position)	ASTM D 2061	15 lbs

F. Appearance

Appearance tests include the change in garment appearance after repeated washings. This includes but is not limited to:
- Good dimensional stability
- Retention of fabric smoothness and recovery properties.
- No chipping, discolouration, or rusting of hardware.
- No fraying or ravelling of buttonholes, trims, or hems.
- No twisting of seams.
- No seam openings or needle cuts.
- No objectionable frosting, pilling, and snagging.
- No deterioration of elasticised areas.
- No deterioration of buttons, snaps or zippers.
- Torque shall be less than 5% of side seam length.
- Differential shrinkage difference shall not be more than 2% between shell and lining.

G. Smoothness appearance

Test	Method	Requirement
Fabric	AATCC 143	Class 4
Seam appearance	AATCC 143	Class 3
Crease retention	AATCC 143	Class 4

2.4.5   Knit slacks, pants, and shorts

I.   *Fabric*

Fabric construction   As approved/contracted ($\pm5\%$)
Flammability   Class 1
(Must comply with 16 CFR part1610 or ASTM D1230)

II.   *Garment construction*

- Pockets – They will be uniform in size and placed evenly or aligned.
- Seams – They must be finished and back tacked at ends. No untrimmed threads are allowed. They must be free from puckering and correct tension shall be used. They must be stretched with fabric without breaking. Non-stretch knits must be taped (unless otherwise specified) or reinforced.
- Stitching – Thread must be colourfast, no broken top stitches, and open seams.
- Darts – They must be uniform in length and shape. There shall be no puckering or bubbles.
- Stress points – They must be bartacked or reinforced as necessary.
- Buttons – They must be securely fastened and colourfast. Button holes must be compatible and completely stitched around.
- Snaps, rivets, and trims – They must be securely fastened, reinforced, and no corrosion after 1 hour at rest in laundry machine after one home laundry cycle.
- Elastic and ribbing – They must extend to fullest width of fabric without breaking stitches. Tunnelled elastic must be stitched down to prevent twisting and rollover. No exposed elastic is allowed.
- Stripes and plaids – They must match at all seams unless otherwise specified.
- Hems and edge finishing – They must be even with no raw edges.
- Needle cutting – Correct needle size and type for fabric are necessary. Ball point needles should be used with knits to prevent needle cutting.

- Zippers – Correct duty zipper for garment is necessary. No bulging or wavering on tape is allowed. Ends of tape, slider, and teeth must be securely fastened.
- Drawstrings – They must be secured/finished at both ends.
- Children garments:There must be no hood or neck drawstrings on garments size 2T-12. Waist/Bottom drawstring on age grades 2T-16 may not exceed 3 inches in length outside the drawstring channel when garment is expanded to its fullest width. No toggles, knots, or attachments at the free end are allowed. Drawstrings must be bartacked at centre back so string cannot be pulled out.
- Pile fabrics – There must be no press marks and crushed pile.
- Painted hardware – It must be less than 0.06% lead by weight.
- Pile fabrics – There must be no press marks and crushed pile.
- Painted hardware – There should be less than 0.06% lead by weight

III. Care labelling

These items must meet all requirements of the Care labelling rule. Provision of practical care instructions is required.

- All care labels must remain legible and securely attached through the laundering cycle and for the life of the garment.
- They will be determined by the testing lab upon completion of all colourways of actual production sample.
- All trims and hardware must be able to withstand selected care method.
- Labels must be bi-lingual (English/Spanish).

IV. Product labelling: The garment must be labelled with

- Brand name
- Fibre content
- Garment size
- Country of origin
- RN# or WPL#; or the name and address of the manufacturer or distributor.

V. Size and fit

- Technical Designer shall verify and approve as per requirement.

VI. Performance standards

A. Fabric performance properties – Appropriate fabric performance standards as mentioned in section 2.3.

B. Seam performance properties

Test	Method	Requirement
Knit and non-woven seams	ASTM D 1683 (Modified)	
Fabric < 3.5 oz/sq yd		30% Elongationor 6 lbs tension
Fabric ≥ 3.5 oz/sq yd		50% Elongationor 7 lbs tension

C. Strength at stress points and of applied decorations

Test	Method	Requirement
Reinforced stress points	ASTM D 1683 (Modified)	20 lbs
Applied decorations	ASTM D 1683 (Modified)	
(i) for 18-month age and up		15 lbs min @ 10 sec
(ii) for under 18-month age		15 lbs. min @ 10 sec
Pocket strength	ASTM D 1683 (Modified)	
	Non-functional	5 lbs
	Functional	10 lbs

D. Snap and button strength properties

Test	Method	Requirement
Anchored strength	ASTM D 1683 (Modified)	
(i) for 18-month age and up and		15 lbs min @10 sec
(ii) for under 18-month age		10 lbs min @10 sec
Button impact resistance	MTL S-1001	5.5 in oz (no failure noted)
Snap/desnap	ASTM D 4846	Opening 2–5 lbs Closing 2–10 lbs

E. Zipper strength properties

Test	Method	Requirement
Seams	ASTM D 2061 (Modified)	20 lbs
Top/bottom stop(open and closed position)	ASTM D 2061	15 lbs

F. Appearance

Appearance tests include the change in garment appearance after repeated washings. This includes but is not limited to:

- Corrosion resistance (metal hardware)   Must display no corrosion
- Lead content of painted hardware   < 0.06%
- Ironing (If recommended).   No colour change
                                               Good appearance
                                               Good shape retention.

  - Good dimensional stability (See appropriate fabric standards).
  - Retention of fabric stretch and recovery properties.
  - No chipping, discolouration, or rusting of hardware.
  - No fraying or ravelling of buttonholes, trims, or hems.
  - No twisting of seams.
  - No seam openings.
  - No objectionable frosting, pilling, and snagging.
  - No deterioration of elasticised areas.
  - No deterioration of buttons, snaps, or zippers.
  - Torque must be less than 5% of side seam length.
  - Differential shrinkage of more than 2% between shell and lining is not allowed.

G. Smoothness appearance

Test	Method	Requirement
Fabric	AATCC 143	Class 4
Seam appearance	AATCC 143	Class 3

2.4.6 Woven dresses, jumpers, rompers, and skirts

I. Fabric

Fabric construction   As approved/contracted ($\pm$5%)
Flammability   Class 1
(Must comply with 16 CFR part1610 or ASTM D1230)

II. Garment construction

- Pockets – They will be uniform in size and placed evenly or aligned.
- Seams – They must be finished and back tacked at ends. No untrimmed threads are allowed. They must be free from puckering, and correct tension shall be used.
- Stitching – Thread must be colourfast, no broken top stitches, and open seams.
- Darts – They must be uniform in length and shape. There shall be no puckering or bubbles.

- Stress points – They must be bartacked or reinforced as necessary.
- Interfacing – They must have compatible shrinkage to shell fabric and must lie flat.
- Lining and attached slips – They must have compatible shrinkage to shell fabric. Linings must lie flat. Linings and slips must be firmly attached and appropriately hemmed.
- Buttons – They must be securely fastened and colourfast. Button holes must be compatible and completely stitched around.
- Snaps, rivets, and trims – They must be securely fastened, reinforced, and no corrosion after 1 hour at rest in laundry machine after one home laundry cycle.
- Elastic and ribbing – They must extend to fullest width of fabric without breaking stitches. Tunnelled elastic must be stitched down to prevent twisting and rollover. No exposed elastic is allowed.
- Stripes and plaids – They must match at all seams unless otherwise specified.
- Plackets –No puckering at seams, especially at bottom or base is desired.
- Hems and edge finishing – They must be even with no raw edges.
- Needle cutting – Correct needle size and type for fabric are required. No needle cuts are expected.
- Zippers – Correct duty zipper for garment is necessary. No bulging or wavering on tape is allowed. Ends of tape must be securely fastened.
- Drawstrings – They must be secured/finished at both ends.
- Children's garments: There must be no hood or neck drawstrings on garments size 2T-12. Waist/Bottom drawstring on age grades 2T-16 may not exceed 3 inches in length outside the drawstring channel when garment is expanded to its fullest width. No toggles, knots, or attachments at the free end are allowed. Drawstrings must be bartacked at centre back so string cannot be pulled out.
- Pile fabrics – There must be no press marks and no crushed pile.
- Painted hardware – They must be less than 0.06% lead by weight.

III. Care labelling

These items must meet all requirements of the care labelling rule. Provision of practical care instructions is required.

- All care labels must remain legible and securely attached through the laundering cycle and for the life of the garment.

- They will be determined by the testing lab upon completion of all colourways of actual production sample.
- All trims and hardware must be able to withstand selected care method.
- Labels must be bi-lingual (English/Spanish).

IV. *Product labelling: The garment must be labelled with*

- Brand name
- Fibre content
- Garment size
- Country of origin
- RN# or WPL#; or the name and address of the manufacturer or distributor.

V. *Size and fit*

- Technical Designer shall verify and approve as per desired specification.

VI. *Performance standards*

A. Fabric performance properties– **A**ppropriate fabric performance standards as mentioned in section 2.3.

B. Seam performance properties

Test	Method	Requirement
Woven seams	ASTM D 1683	
Fabric < 3.5 oz/sq yd		15 lbs/in
Fabric ≥ 3.5 oz/sq yd		20 lbs/in

C. Strength at stress points and of applied decorations

Test	Method	Requirement
Reinforced stress points	ASTM D 1683 (Modified)	20 lbs
Applied decorations	ASTM D 1683 (Modified)	
(i) for 18-month age and up		15 lbs min @ 10 sec
(ii) for under 18-month age		10 lbs. min @ 10 sec
Pocket strength	ASTM D 1683 (Modified)	
Under 8 oz/sq yd	Non-functional	10 lbs
	Functional	15 lbs
Over 8 oz/sq yd	Non-functional	10 lbs
	Functional	20 lbs

D. Button and snap strength properties

Test	Method	Requirement
Anchored strength (i) for 18-month age and up and (ii) for under 18-month age	ASTM D 1683 (Modified)	15 lbs min @10 sec 10 lbs min @10 sec
Button impact resistance	MTL S-1001	5.5 in oz (no failure noted)
Snap/desnap	ASTM D 4846	Opening 2–5 lbs Closing2–10 lbs

E. Zipper strength properties

Test	Method	Requirement
Seams	ASTM D 2061 (Modified)	20 lbs
Top/bottom stop (open and closed position)		15 lbs

F. Appearance

Appearance tests include the change in garment appearance after repeated washings. This includes but is not limited to:

- Corrosion resistance (metal hardware).   Must display no corrosion
- Lead content of painted hardware   < 0.06%
- Ironing (If recommended)   No colour change
    Good appearance
    Good shape retention.

    - Good dimensional stability.
    - Retention of fabric smoothness and recovery properties.
    - No chipping, discolouration, or rusting of hardware.
    - No fraying or ravelling of buttonholes, trims, or hems.
    - No twisting of seams.
    - No seam openings or needle cuts.
    - No objectionable frosting, pilling, and snagging.
    - No deterioration of elasticised areas.
    - No deterioration of buttons, snaps, or zippers.
    - Torque shall be less than 5% of side seam length.
    - Differential shrinkage difference shall not be more than 2% difference between shell and lining.

G. Smoothness appearance

Test	Method	Requirement
Fabric	AATCC 143	Class 4
Seam appearance	AATCC 143	Class 3

2.4.7   Knit dresses, jumpers, rompers, and skirts

I. Fabric

Fabric construction   As approved/contracted ($\pm$5%)
Flammability   Class 1
(Must comply with 16 CFR part1610 or ASTM D1230)

II. Garment construction

- Pockets – They will be uniform in size and placed evenly or aligned.
- Seams – They must be finished and back tacked at ends. No untrimmed threads are allowed. They must be free from puckering appropriate tension, and seam type shall be used. They must stretch with fabric without breaking.
- Shoulder seams – They must be taped or reinforced unless otherwise specified.
- Stitching – Thread must be colourfast. No broken top stitches, open seams, and breakage are allowed when fabric is fully extended.
- Darts – They must be uniform in length and shape and no puckering or bubbles.
- Stress points – They must be bartacked or reinforced as necessary.
- Interfacing – They must have compatible shrinkage to shell fabric and must lie flat.
- Lining and attached slips – They must have compatible shrinkage to shell fabric. Linings must lie flat. Linings and slips must be firmly attached and appropriately hemmed.
- Buttons – They must be securely fastened and colourfast. Button holes must be compatible and completely stitched around.
- Snaps, rivets, and trims – They must be securely fastened, reinforced, and no corrosion after 1 hour at rest in laundry machine after one home laundry cycle.
- Elastic and ribbing – They must extend to fullest width of fabric without breaking stitches. Tunnelled elastic must be stitched down to prevent twisting and rollover. No exposed elastic is allowed.
- Stripes and plaids – They must match at all seams unless otherwise specified.

- Plackets – No puckering at seams is allowed, especially at bottom or base.
- Hems and edge finishing – They must be even with no raw edges.
- Needle cutting – Correct needle size and type for fabric are necessary. Ball point needles should be used with knits to prevent needle cutting.
- Zippers– Correct duty zipper for garment is necessary. No bulging or wavering on tape is allowed. Ends of tape must be securely fastened.
- Drawstrings – They must be secured/finished at both ends.
- Children garments: There must be no hood or neck drawstrings on garments size 2T-12. Waist/Bottom drawstring on age grades 2T-16 may not exceed 3 inches in length outside the drawstring channel when garment is expanded to its fullest width. No toggles, knots, or attachments at the free end are allowed. Drawstrings must be bartacked at centre back so string cannot be pulled out.
- Pile fabrics – There must be no press marks and no crushed pile.
- Painted hardware – There should be less than 0.06% lead by weight

III. Care labelling

These items must meet all requirements of the Care labelling rule. Provision of practical care instructions is required.

- All care labels must remain legible and securely attached through the laundering cycle and for the life of the garment.
- They will be determined by the testing lab upon completion of all colourways of actual production sample.
- All trims and hardware must be able to withstand selected care method
- Labels must be bi-lingual (English/Spanish).

IV. Product labelling

The garment must be labelled with:

- Brand name
- Fibre content
- Garment size
- Country of origin
- RN# or WPL#; or the name and address of the manufacturer or distributor.

V. Size and fit

- Technical Designer shall verify and approve as per size specification.

VI. Performance standards

A. Fabric performance properties – Appropriate fabric performance standards as mentioned in section 2.3.

B. Seam performance properties

Test	Method	Requirement
Knit and non-woven seams	ASTM D 1683 (Modified)	
Fabric < 3.5 oz/sq yd		30% Elongationor 6 lbs tension
Fabric ≥ 3.5 oz/sq yd		50% Elongationor 7 lbs tension

C. Strength at stress points and of applied decorations

Test	Method	Requirement
Reinforced stress points	ASTM D 1683 (Modified)	20 lbs
Applied decorations	ASTM D 1683 (Modified)	
(i) for 18-month age and up		15 lbs min @ 10 sec
(ii) for under 18-month age		10 lbs. min @ 10 sec
Pocket strength	ASTM D 1683(Modified) Non-functional Functional	5 lbs 10 lbs

D. Snap and button strength properties

Test	Method	Requirement
Anchored strength	ASTM D 1683(Modified)	
(i) for 18-month age and up and		15 lbs min @10 sec
(ii) for under 18-month age		10 lbs min @10 sec
Button impact resistance	MTL S-1001	5.5 in oz (no failure noted)
Snap/desnap	ASTM D 4846	Opening 2–5 lbs Closing 2–10 lbs

E. Zipper strength properties

Test	Method	Requirement
Seams	ASTM D 2061(Modified)	20 lbs
Top/bottom stop(open and closed position)	ASTM D 2061	15 lbs

F. Appearance

Appearance tests include the change in garment appearance after repeated washings. This includes but is not limited to:

- Corrosion resistance (metal hardware) Must display no corrosion
- Lead content of painted hardware < 0.06%
- Ironing (If recommended) No colour change
  Good appearance
  Good shape retention.

  - Good dimensional stability.
  - Retention of fabric stretch and recovery properties.
  - No chipping, discolouration, or rusting of hardware.
  - No fraying or ravelling of buttonholes, trims, or hems.
  - No twisting of seams.
  - No seam openings.
  - No objectionable frosting, pilling, and snagging.
  - No deterioration of elasticised areas.
  - No deterioration of buttons, snaps, or zippers.
  - Torque shall be less than 5% of side seam length.
  - Differential shrinkage difference shall not be more than 2% between shell and lining.

2.4.8   Activewear

I. Fabric

Fabric construction As approved/contracted (±5%)
Flammability Class 1
(Must comply with 16 CFR part1610 or ASTM D1230 required)

II.   Garment construction

- Pockets – They will be uniform in size and placed evenly or aligned.
- Seams – They must be finished and back tacked at ends. No untrimmed threads are allowed. They must be free from puckering appropriate tension, and seam type shall be used. They must stretch with fabric without breaking.

- Shoulder seams – They must be taped or reinforced unless otherwise specified.
- Stitching– Thread must be colourfast. No broken top stitches, open seams, and breakage are allowed when fabric is fully extended.
- Darts– They must be uniform in length and shape and no puckering or bubbles.
- Stress points – They must be bartacked or reinforced as necessary.
- Interfacing – They must have compatible shrinkage to shell fabric and must lie flat.
- Buttons– They must be securely fastened and colourfast. Button holes must be compatible and completely stitched around.
- Snaps, rivets, and trims – They must be securely fastened, reinforced, and no corrosion after 1 hour at rest in laundry machine after one home laundry cycle.
- Elastic and ribbing – They must extend to fullest width of fabric without breaking stitches.Tunnelled elastic must be stitched down to prevent twisting and rollover. No exposed elastic is allowed.
- Stripes and plaids – They must match at all seams unless otherwise specified.
- Plackets – No puckcring at seams, especially at bottom or base is desired.
- Hems and edge finishing – They must be even with no raw edges.
- Needle cutting – Correct needle size and type for fabric are required. No needle cuts are expected.
- Spandex – There must be no exposed or cut strands of spandex yarns.
- Zippers – Correct duty zipper for garment is necessary. No bulging or wavering on tape is allowed. Ends of tape must be securely fastened.
- Drawstrings – They must be secured/finished at both ends.
  Children's garments: There must be no hood or neck drawstrings on garments size 2T-12. Waist/Bottom drawstring on age grades 2T-16 may not exceed 3 inches in length outside the drawstring channel when garment is expanded to its fullest width.No toggles, knots, or attachments at the free end are allowed. Drawstrings must be bartacked at centre back so string cannot be pulled out.
- Pile fabrics– There must be no press marks and no crushed pile.
- Painted hardware – There should be less than 0.06% lead by weight

III. *Care labelling:*

These items must meet all requirements of the Care labelling rule. Provision of practical care instructions is required.

- All care labels must remain legible and securely attached through the laundering cycle and for the life of the garment.
- They will be determined by the testing lab upon completion of all colourways of actual production sample.
- All trims and hardware must be able to withstand selected care method.
- Labels must be bi-lingual (English/Spanish)

IV. Product labelling:The garment must be labelled with:

- Brand name
- Fibre content
- Garment size
- Country of origin
- RN# or WPL#; or the name and address of the manufacturer or distributor.

V. Size and fit

- Technical Designer shall verify and approve as per required specification.

VI. Performance standards

A. Fabric performance properties – Appropriate fabric performance standards as mentioned in section 2.3.

B. Seam performance properties

Test	Method	Requirement
Woven seams	ASTM D 1683	
Fabric <3.5 oz/sq yd		15 lbs./in
Fabric ≥3.5 oz/sq yd		20 lbs./in
Knit and non-woven seams	ASTM D 1683 (Modified)	
Fabric < 3.5 oz/sq yd		30% Elongationor 6 lbs tension
Fabric ≥ 3.5 oz/sq yd		50% Elongationor 7 lbs tension
Seam slippage	ASTM D434[14]	15 lbs./in

C. Strength at stress points and of applied decorations

Test	Method	Requirement
Reinforced stress points	ASTM D 1683 (Modified)	20 lbs
Applied decorations	ASTM D 1683 (Modified)	
(i) for 18-month age and up		15 lbs min @ 10 sec
(ii) for under 18-month age		10 lbs. min @ 10 sec
Woven pocket strength	ASTM D 683 (Modified)	
(i) Under 8 oz/sq yd	Non-functional	10 lbs
	Functional	15 lbs
(ii) Over 8 oz/sq yd	Non-functional	10 lbs
	Functional	20 lbs
Knit pocket strength	ASTM D 683 (Modified)	
	Non-functional	5 lbs
	Functional	10 lbs

D. Button and snap strength properties

Test	Method	Requirement
Anchored strength	ASTM D 1683 (Modified)	
(i) for 18-month age and up and		15 lbs min @10 sec
(ii) for under 18-month age		10 lbs min @10 sec
Button impact resistance	MTL S-1001	5.5 in oz (no failure noted)
Snap/desnap	ASTM D 4846	Opening 2–5 lbs Closing 2–10 lbs

E. Zipper strength properties

Test	Method	Requirement
Seams	ASTM D 2061	20 lbs
Top/bottom stop (open and closed position)	ASTM D 2061 (Modified)	15 lbs

F. Water repellency (if claimed)

Test	Method	Requirement
Original state	AATCC 22[15]	90 (ISO-4) -(1) ISO Class
After one wash (if permanent)		70 (ISO-2)-(1) ISO Class

G. Water resistance (if claimed)

Test	Method	Requirement
Original state	AATCC 35[16]	1 g max.

H. Appearance

Appearance tests include the change in garment appearance after repeated washings. This includes but is not limited to:

- Corrosion resistance (metal hardware) Must display no corrosion
- Lead content of painted hardware < 0.06%
- Ironing (If recommended) No colour change
  Good appearance
  Good shape retention.

  - Good dimensional stability (See appropriate fabric standards).
  - Retention of fabric stretch and recovery properties.
  - No chipping, discolouration, or rusting of hardware.
  - No fraying or ravelling of buttonholes, trims, or hems.
  - No twisting of seams.
  - No seam openings.
  - No objectionable frosting, pilling, and snagging.
  - No deterioration of elasticised areas.
  - No deterioration of buttons, snaps, or zippers.
  - Torque shall be less than 5% of side seam length.
  - Differential shrinkage difference shall not be more than 2% between shell and lining.

2.4.9 Sleepwear

I. Fabric

Fabric construction As approved/contracted (±5%)
Flammability

- Adult sleepwear Must comply with 16 CFR part1610 or ASTM D1230

- Children's sleepwear   Up to Sizes 6X/7– must comply with 16 CFR parts 1615 and 1616[17]

II. *Garment construction*

- Pockets – They will be uniform in size and placed evenly or aligned.
- Seams – They must be finished and back tacked at ends. No untrimmed threads are allowed. They must be free from puckering, and correct tension shall be used. They must be stretched with fabric without breaking. Non-stretch knits must be taped (unless otherwise specified) or reinforced.
- Stitching – Thread must be colourfast, no broken top stitches, and open seams.
- Darts – They must be uniform in length and shape. There shall be no puckering or bubbles.
- Stress points – They must be bartacked or reinforced as necessary
- Buttons – They must be securely fastened and colourfast. Button holes must be compatible and completely stitched around.
- Snaps, rivets, and trims –They must be securely fastened, reinforced, and no corrosion after 1 hour at rest in laundry machine after one home laundry cycle.
- Elastic and ribbing – They must extend to fullest width of fabric without breaking stitches. Tunnelled elastic must be stitched down to prevent twisting and rollover. No exposed elastic is allowed.
- Stripes and plaids – They must match at all seams unless otherwise specified.
- Plackets – There must be no puckering at seams, especially at bottom or base.
- Hems and edge finishing – They must be even with no raw edges.
- Needle cutting – Correct needle size and type for fabric are necessary. Ball point needles should be used with knits to prevent needle cutting.
- Zippers – Correct duty zipper for garment is necessary. No bulging or wavering on tape is allowed.Ends of tape,slider, and teeth must be securely fastened.
- Drawstrings – They must be secured/finished at both ends.
- Children's garments: There must be no hood or neck drawstrings on garments size 2T-12. Waist/Bottom drawstring on age grades 2T-16 may not exceed 3 inches in length outside the drawstring channel when garment is expanded to its fullest width. No toggles, knots, or attachments at the free end are allowed. Drawstrings must be bartacked at centre back so string cannot be pulled out.

- Pile fabrics – There must be no press marks and crushed pile.
- Painted hardware – It must be less than 0.06% lead by weight.

III.   Care labelling:

These items must meet all requirements of the care labelling rule.Provision of practical care instructions is required.

- All care labels must remain legible and securely attached through the laundering cycle and for the life of the garment.
- Will be determined by the testing lab upon completion of all colourways of actual production sample.
- All trims and hardware must be able to withstand selected care method.
- Labels must be bi-lingual (English/Spanish)

IV.   Product labelling: The garment must be labelled with:

- Brand name
- Fibre content
- Garment size
- Country of origin
- RN# or WPL#; or the name and address of the manufacturer or distributor.

VI.   Size and fit

- Technical Designer shall verify and approve as per desired specification.

VI.   Performance standards

A. Fabric performance properties – Appropriate fabric performance standards as mentioned in section 2.3.

B. Seam performance properties

Test	Method	Requirement
Woven seams	ASTM D 1683	
Fabric < 3.5 oz/sq yd		15 lbs./in
Fabric ≥ 3.5 oz/sq yd		20 lbs./in
Knitand non-woven seams	ASTM D 1683 (Modified)	
Fabric < 3.5 oz/sq yd		30% Elongationor 6 lbs tension
Fabric≥ 3.5 oz/sq yd		50% Elongationor 7 lbs tension
Seam slippage	ASTM D434	15 lbs./in

Children's sleepwear – all seam types must pass flammability testing.

C. Strength at stress points and of applied decorations

Test	Method	Requirement
Reinforced stress points	ASTM D 1683 (Modified)	20 lbs
Applied decorations	ASTM D 683 (Modified)	
(i) for 18-month age and up		15 lbs min @ 10 sec
(ii) for under 18-month age		10 lbs. min @ 10 sec
Woven pocket strength	ASTM D 683 (Modified)	
(i) Under 8 oz/sq yd	Non-functional	10 lbs
	Functional	15 lbs
(ii) Over 8 oz/sq yd	Non-functional	10 lbs
	Functional	20 lbs
Knit pocket strength	ASTM D 683 (Modified)	
	Non-functional	5 lbs
	Functional	10 lbs

D. Button and snap strength properties

Test	Method	Requirement
Anchored strength	ASTM D 1683	
(i) for 18-month age and up and	(Modified)	15 lbs min @10 sec
(ii) for under 18-month age		10 lbs min @10 sec
Button impact resistance	MTL S-1001	5.5 in oz (no failure noted)
Snap/desnap	ASTM D 4846	Opening 2–5 lbs Closing 2–10 lbs

E. Zipper strength properties

Test	Method	Requirement
Seams	ASTM D 2061	20 lbs
Top/bottom stop (open and closed position)	ASTM D 2061 (Modified)	15 lbs

F. Appearance

Appearance tests include the change in garment appearance after repeated washings. This includes but is not limited to:

- Corrosion resistance (metal hardware)   Must display no corrosion
- Lead content of painted hardware   < 0.06%
- Ironing (If recommended)   No colour change

Good appearance
Good shape retention.

- Good dimensional stability.
- Retention of fabric stretch and recovery properties.
- No chipping, discolouration, or rusting of hardware.
- No fraying or ravelling of buttonholes, trims, or hems.
- No twisting of seams.
- No seam openings.
- No objectionable frosting, pilling, and snagging.
- No deterioration of elasticised areas.
- No deterioration of buttons, snaps, or zippers.
- Torque shall be less than 5% of side seam length.
- Differential shrinkage difference shall not be more than 2% between shell and lining.

2.4.10   Underwear – panties, briefs, and boxer shorts

I.   Fabric

Fabric construction   As approved/contracted ($\pm$5%)
    Flammability   Class 1
    (Must comply with 16 CFR part1610 or ASTM D1230)
    Children's 2 pc underwear   Must be labelled/tagged as "not intended for use as sleepwear."

II.   Garment construction

- Seams – They must be finished and back tacked at ends. No untrimmed threads are allowed. They must be free from puckering, and correct tension shall be used. They must be stretched with fabric without breaking. Non-stretch knits must be taped (unless otherwise specified) or reinforced.
- Stitching– Thread must be colourfast, no broken top stitches and open seams.
- Stress points – They must be bartacked or reinforced as necessary.
- Trims and lace – They must be firmly attached and must have compatible shrinkage to base fabrics.
- Appliqués – They must be firmly secured with no exposed raw edges. They must have compatible shrinkage to base fabric.
- Ribbons and bows – They must be firmly attached. Raw ends of bows must be heat sealed or hemmed to prevent ravelling.

- Buttons – They must be securely fastened and colourfast. Button holes must be compatible and completely stitched around.
- Snaps, rivets, and trims – They must be securely fastened, reinforced, and no corrosion after 1 hour at rest in laundry machine after one home laundry cycle.
- Elastic and ribbing– They must extend to fullest width of fabric without breaking stitches. Tunnelled elastic must be stitched down to prevent twisting and rollover. No exposed elastic is allowed.
- Stripes and plaids – They must match at all seams unless otherwise specified.
- Plackets – There must be no puckering at seams, especially at bottom or base.
- Drawstrings – They must be secured/finished at both ends.
- Children's garments:There must be no hood or neck drawstrings on garments size 2T-12. Waist/Bottom drawstring on age grades 2T-16 may not exceed 3 inches in length outside the drawstring channel when garment is expanded to its fullest width.No toggles, knots, or attachments at the free end are allowed. Drawstrings must be bartacked at centre back so string cannot be pulled out.
- Hems and edge finishing – There must be even with no raw edges.
- Needle cutting – Correct needle size and type for fabric shall be used. Ball point needles should be used with knits to prevent needle cutting.
- Napped fabrics – There must be no press marks and no crushed nap.
- Painted hardware – There must be less than 0.06 % lead by weight.

III. *Care labelling:*

These items must meet all requirements of the care labelling rule. Provision of practical care instructions is required.

- All care labels must remain legible and securely attached through the laundering cycle and for the life of the garment.
- They will be determined by the testing lab upon completion of all colourways of actual production sample.
- All trims and hardware must be able to withstand selected care method.
- Labels must be bi-lingual (English/Spanish)

IV. Product labelling: The garment must be labelled with:

- Brand name
- Fibre content
- Garment size
- Country of origin
- RN# or WPL#; or the name and address of the manufacturer or distributor.

V. Size and fit

- Technical Designer shall verify and approve as per desired specification.

VI. Performance standards

A. Fabric performance properties – Appropriate fabric performance standards as mentioned in section 2.3.

B. Seam performance properties

Test	Method	Requirement
Woven seams	ASTM D 1683	
Fabric < 3.5 oz/sq yd		15 lbs./in
Fabric ≥ 3.5 oz/sq yd		20 lbs./in
Knit and non-woven seams	ASTM D 1683 Modified)	
Fabric < 3.5 oz/sq yd		30% Elongationor 6 lbs tension
Fabric ≥ 3.5 oz/sq yd		50% Elongationor 7 lbs tension
Seam slippage	ASTM D434	15 lbs./in

C. Strength at stress points and of applied decorations

Test	Method	Requirement
Reinforced stress points	ASTM D 1683 (Modified)	20 lbs
Applied decorations	ASTM D 1683 (Modified)	
(i) for 18-month age and up		15 lbs min @ 10 sec
(ii) for under 18-month age		10 lbs. min @ 10 sec

(Contd.)

Test	Method	Requirement
Woven pocket strength	ASTM D 683 (Modified)	
(i)Under 8 oz/sq yd	Non-functional	10 lbs
	Functional	15 lbs
(ii) Over 8 oz/sq yd	Non-functional	10 lbs
	Functional	20 lbs
Knit pocket strength	ASTM D 1683 (Modified)	
	Non-functional	5 lbs
	Functional	10 lbs

D. Button and snap strength properties

Test	Method	Requirement
Anchored strength	ASTM D 683 (Modified)	
(i) for 18-month age and up and		15 lbs min @10 sec
(ii) for under 18-month age		10 lbs min @10 sec
Button impact resistance	MTL S-1001	5.5 in oz (no failure noted)
Snap/desnap	ASTM D 4846	Opening 2–5 lbs Closing 2–10 lbs

E. Zipper strength properties

Test	Method	Requirement
Seams	ASTM D 2061	20 lbs
Top/bottom stop (open and closed position)	ASTM D 2061 (Modified)	15 lbs

F. Appearance

Appearance tests include the change in garment appearance after repeated washings. This includes but is not limited to:

- • Corrosion resistance (metal hardware)   Must display no corrosion
- • Lead content of painted hardware   < 0.06%
- • Ironing (If recommended)   No colour change
  Good appearance
  Good shape retention.
  - – Good dimensional stability.
  - – Retention of fabric stretch and recovery properties.

- No chipping, discolouration, or rusting of hardware.
- No fraying or ravelling of buttonholes, trims, or hems.
- No twisting of seams.
- No seam openings.
- No objectionable frosting, pilling, and snagging.
- No deterioration of elasticised areas.
- No deterioration of buttons, snaps, or zippers.
- Torque shall be less than 5% of side seam length.
- Differential shrinkage difference shall not be more than 2% between shell and lining.

2.4.11   Underwear – brassieres and foundation garments

I.   Fabric

Fabric construction   As approved/contracted ($\pm$5%)
Flammability   Class 1
(Must comply with 16 CFR part1610 or ASTM D1230)

II.   Garment construction

- Seams – They must be finished and back tacked at ends. No untrimmed threads are allowed. They must be free from puckering and correct tension shall be used. They must be stretched with fabric without breaking. Non-stretch knits must be taped (unless otherwise specified) or reinforced.
- Stitching – Thread must be colourfast, no broken top stitches, and open seams.
- Stress points – They must be bartacked or reinforced as necessary.
- Linings and interlinings – They must have compatible shrinkage to shell fabric. Interlining must remain secure, not shift, and not migrate.
- Trims and lace   There must be firmly attached and have compatible shrinkage to base fabrics.
- Removable bra pads, "cookies"–There must be no open seams. Interlining must remain secure, not shift, and not migrate. Raw ends must be overcast or hemmed to prevent ravelling.
- Appliqués, ribbons and bows – They must be firmly attached. Raw ends of bows must be heat sealed or hemmed to prevent ravelling.
- Underwire – They must be securely fastened;completely covered;remain secure; and not corrode, break, distort, or burst through support fabric during or following laundering. There shall be no corrosion after 1 hour at rest in laundry machine after one home laundry cycle.

- Hooks, sliders, and other metal or plastic closures–No corroding or chipping is allowed and must be securely fastened and reinforced. They must not break or distort during laundering and no corrosion after 1 hour at rest in laundry machine after one home laundry cycle.
- Elastic and ribbing – They must extend to fullest width of fabric without breaking stitches.Tunnelled elastic must be stitched down to prevent twisting and rollover. No exposed elastic shall be present unless otherwise specified.
- Hems and edge finishing – They must be even with no raw edges.
- Needle cutting–Correct needle size and type for fabric must be used. Ball point needles should be used with knits to prevent needle cutting.
- Napped fabrics – There must be no press marks and no crushed nap.
- Painted hardware – There must be less than 0.06% lead by weight.

III. Care labelling:

These items must meet all requirements of the care labelling rule. Provision of practical care instructions is required.

- All care labels must remain legible and securely attached through the laundering cycle and for the life of the garment.
- They will be approved by the testing lab upon completion of all colourways of actual production sample.
- All trims and hardware must be able to withstand selected care method.
- Labels must be bi-lingual (English/Spanish)

IV. Product labelling:The garment must be labelled with:

- Brand name
- Fibre content
- Garment size
- Country of origin
- RN# or WPL#; or the name and address of the manufacturer or distributor.

V. Size and fit

- Technical Designer shall verify and approve as per desired specification.

VI. Performance standards

A. Fabric performance properties – Appropriate fabric performance standards as mentioned in section 2.3.

B. Seam performance properties

Test	Method	Requirement
Woven seams	ASTM D 1683	
Fabric < 3.5 oz/sq yd		15 lbs./in
Fabric ≥ 3.5 oz/sq yd		20 lbs./in
Knitand non-woven seams	ASTM D 1683 (Modified)	
Fabric < 3.5 oz/sq yd		30% Elongationor 6 lbs tension
Fabric ≥ 3.5 oz/sq yd		50% Elongationor 7 lbs tension
Seam slippage	ASTM D434	15 lbs./in

C. Strength at stress points and of applied decorations

Test	Method	Requirement
Reinforced stress points	ASTM D 1683 (Modified)	20 lbs
Applied decorations	ASTM D 1683 (Modified)	
(i) for 18-month age and up		15 lbs min @ 10 sec
(ii) for under 18-month age		10 lbs. min @ 10 sec
Woven pocket strength	ASTM D 1683 (Modified)	
(i) Under 8 oz/sq yd	Non-functional	10 lbs
	Functional	15 lbs
(ii) Over 8 oz/sq yd	Non-functional	10 lbs
	Functional	20 lbs
Knit pocket strength	ASTM D 1683 (Modified)	
	Non-functional	5 lbs
	Functional	10 lbs

D. Button and snap strength properties

Test	Method	Requirement
Anchored strength	ASTM D 1683 (Modified)	
(i) for 18-month age and up and		15 lbs min @10 sec
(ii) for under 18-monthage		10 lbs min @10 sec
Button impact resistance	MTL S-1001	5.5 in oz (no failure noted)
Snap/desnap	ASTM D 4846	Opening 2–5 lbs Closing 2–10 lbs

E. Zipper strength properties

Test	Method	Requirement
Seams	ASTM D 2061	20 lbs
Top/bottom stop (open and closed position)	ASTM D 2061 (Modified)	15 lbs

F. Appearance

Appearance tests include the change in garment appearance after repeated washings. This includes but is not limited to:

- Corrosion resistance (metal hardware).   Must display no corrosion
- Lead content of painted hardware   < 0.06%
- Ironing (If recommended)   No colour change
                                            Good appearance
                                            Good shape retention.

  - Good dimensional stability.
  - Retention of fabric stretch and recovery properties.
  - No chipping, discolouration, or rusting of hardware.
  - No distortion or damage of underwire or hardware
  - No fraying or ravelling of laces, trims, or hems.
  - No twisting of seams.
  - No seam openings.
  - No objectionable frosting, pilling, and snagging
  - No fibre migration or separation of fused materials.
  - No deterioration of elasticised areas.
  - Torque shall be less than 5% of side seam length.
  - Differential shrinkage difference shall not be more than 2% between shell and lining.

2.4.12 Outerwear

I. Fabric

Fabric construction   As approved/contracted ($\pm$5%)
Flammability   Class 1
(Must comply with 16 CFR part1610 or ASTM D1230)

II. Garment construction

- Pockets – They will be uniform in size and placed evenly or aligned.
- Seams – They must be finished and back tacked at ends. No untrimmed threads are allowed. They must be free from puckering appropriate

tension, and seam type shall be used. They must stretch with fabric without breaking.

- Shoulder seams – They must be taped or reinforced unless otherwise specified.
- Stitching – Thread must be colourfast. No broken top stitches, open seams, and breakage are allowed when fabric is fully extended.
- Darts and vents – They must be uniform in length and shape. No puckering or bubbles is allowed.
- Stress points – They must be bartacked or reinforced as necessary.
- Interfacing – They must have compatible shrinkage to shell fabric and must lie flat.
- Trims – They must be securely fastened, colourfast, and have shrinkage similar to shell fabric.
- Buttons – They must be securely fastened and colourfast. Button holes must be compatible and completely stitched around.
- Snaps, rivets, toggles, and rigid trims – No corroding, securely fastened, and reinforced. No chipping or flaking of paint. No corrosion after 1 hour at rest in laundry machine after one home laundry cycle.
- Painted hardware – There must be less than 0.06% lead by weight.
- Elastic and ribbing – They must extend to fullest width of fabric without breaking stitches. Tunnelled elastic must be stitched down to prevent twisting and rollover. No exposed elastic is allowed.
- Stripes and plaids – They must match at all seams unless otherwise specified.
- Plackets – There shall be no puckering at seams, especially at bottom or base.
- Hems and edge finishing – They must be even with no raw edges.
- Needle cutting – Correct needle size and type for fabric shall be used. Ball point needles mustbe used with knits to prevent needle cutting.
- Spandex – No exposed or cut strands of spandex yarns is allowed.
- Zippers – They must be correct duty zipper for garment. No bulging or wavering on tape is allowed. Ends of tape must be securely fastened. Slide must move easily when very cold/icy.
- Drawstrings – They must be secured/finished at both ends.
- Children's garments:There must be no hood or neck drawstrings on garments size 2T-12. Waist/Bottom drawstring on age grades 2T-16 may not exceed 3 inches in length outside the drawstring channel when garment is expanded to its fullest width. No toggles, knots, or attachments at the free end are allowed. Drawstrings must be bartacked at centre back so string cannot be pulled out.

- Pile fabrics – There should be no press marks and no crushed pile.
- Interlining – They must be evenly distributed, no lumps/clumps and no migration through shell or lining fabric.

III.   Care labelling

These items must meet all requirements of the care labelling rule. Provision of practical care instructions is required.

- All care labels must remain legible and securely attached through the laundering cycle and for the life of the garment.
- They will be determined by the testing lab upon completion of all colourways of actual production sample.
- All trims and hardware must be able to withstand selected care method
- Labels must be bi-lingual (English/Spanish)

IV.   Product labelling: The garment must be labelled with:

- Brand name
- Fibre content
- Garment size
- Country of origin
- RN# or WPL#; or the name and address of the manufacturer or distributor.

V.   Size and fit

- Technical Designer shall verify and approve as per desired specification.

VI.   Performance standards

A. Fabric performance properties –Appropriate fabric performance standards as mentioned in section 2.3.

B. Seam performance properties

Test	Method	Requirement
Woven seams	ASTM D 1683	
Fabric < 3.5 oz/sq yd		15 lbs./in
Fabric ≥ 3.5 oz/sq yd		20 lbs./in
Knitand non-woven seams	ASTM D 1683 (Modified)	
Fabric < 3.5 oz/sq yd		30% Elongationor 6 lbs tension
Fabric ≥ 3.5 oz/sq yd		50% Elongationor 7 lbs tension
Seam slippage	ASTM D434	15 lbs./in

C. Strength at stress points and of applied decorations

Test	Method	Requirement
Reinforced stress points	ASTM D 1683 (Modified)	20 lbs
Applied decorations	ASTM D 1683 (Modified)	
(i) for 18-month age and up		15 lbs min @ 10 sec
(ii) for under 18-month age		10 lbs. min @ 10 sec
Woven pocket strength	ASTM D 1683 (Modified)	
(i) Under 8 oz/sq yd	Non-functional	10 lbs
	Functional	15 lbs
(ii) Over 8 oz/sq yd	Non-functional	10 lbs
	Functional	20 lbs
Knit pocket strength	ASTM D 1683 (Modified)	
	Non-functional	5 lbs
	Functional	10 lbs

D. Button and snap strength properties

Test	Method	Requirement
Anchored strength	ASTM D 1683 (Modified)	
(i) for 18-monthage and up and		15 lbs min @10 sec
(ii) for under 18-month age		10 lbs min @10 sec
Button impact resistance	MTL S-1001	5.5 in oz (no failure noted)
Snap/Desnap	ASTM D 4846	Opening 2-5 lbs Closing 2-10 lbs

E. Zipper strength properties

Test	Method	Requirement
Seams	ASTM D 2061	20 lbs
Top/bottom stop(open and closed position)	ASTM D 2061(Modified)	15 lbs

F. Water repellency (if claimed)

Test	Method	Requirement
Original state	AATCC 22	90 (ISO-4)-(1) ISO Class
After 1 wash (if permanent)		70 (ISO-2)-(1) ISO Class

G. Water resistance (if claimed)

Test	Method	Requirement
Original state unless otherwise specified	AATCC 35	1 gram max.

H. Properties of interlining materials – battings/fillers

Test	Method	Requirement
Manmade fibre batting	ASTM D 4770[18]	Class 3.5
Down and down blends	ASTM D 4524 (19) /4522[20]	As specified

I. Appearance

Appearance tests include the change in garment appearance after repeated washings.This includes, but is not limited to:

- Corrosion resistance (metal hardware)   Must display no corrosion
- Lead content of painted hardware   < 0.06%
- Ironing (If recommended)   No colour change
                                                 Good appearance
                                                 Good shape retention.

- Good dimensional stability.
- Retention of fabric stretch and recovery properties.
- No chipping, discolouration or rusting of hardware.
- No fraying or ravelling of buttonholes, trims or hems.
- No twisting of seams.-No seam openings.
- No objectionable frosting, pilling and snagging.
- No deterioration of elasticised areas.
- No deterioration of buttons, snaps or zippers.
- Batting/interlinings/fillers shall retain their integrity following laundering and wear.
- Torque shall be less than 5% of side seam length.
- Differential shrinkage difference shall not be more than 2% between shell and lining.

2.4.13   Swimwear

I. Fabric

Fabric construction	As approved/contracted ($\pm$5%)
Flammability	Class 1
(Must comply with 16 CFR part1610 or ASTM D1230)	

II. *Garment construction*

- Pockets – They will be uniform in size and placed evenly or aligned.
- Seams – They must be finished and back tacked at ends. No untrimmed threads are allowed. They must be free from puckering appropriate tension, and seam type shall be used. They must stretch with fabric without breaking.
- Shoulder seams – They must be taped or reinforced unless otherwise specified.
- Stitching – Thread must be colourfast. No broken top stitches, open seams, and breakage are allowed when fabric is fully extended.
- Darts and pleats – They must be uniform in length and shape unless otherwise specified and no puckering or bubbles is allowed.
- Stress points – They must be bartacked or reinforced as necessary.
- Linings and interlinings – They must have compatible shrinkage to shell fabric. Interlining must remain secure, not shift, and not migrate. They must provide opacity where specified.
- Trims and lace – They must be firmly attached and must have compatible shrinkage to base fabrics.
- Removable bra pads, "cookies" – No open seams are allowed. Interlining must remain secure, not shift, and not migrate.Raw ends must be overcast or hemmed to prevent ravelling.
- Appliqués, ribbons and bows – They must be firmly attached. Raw ends of bows must be heat sealed or hemmed to prevent ravelling.
- Underwire – They must be securely fastened;completely covered;remain secure; and not corrode, break, distort, or burst through support fabric during or following laundering. There shall be no corrosion after 1 hour at rest in laundry machine after one home laundry cycle.
- Hooks, sliders, and other metal or plastic closures – No corroding or chipping is allowed and must be securely fastened and reinforced. They must not break or distort during laundering and no corrosion after 1 hour at rest in laundry machine after one home laundry cycle.
- Snaps, rivets, toggles, and rigid trims – No corroding, securely fastened, and reinforced. No chipping or flaking of paint. No corrosion after 1 hour at rest in laundry machine after one home laundry cycle.
- Elastic and ribbing – They must extend to fullest width of fabric without breaking stitches.Tunnelled elastic must be stitched down to prevent twisting and rollover. No exposed elastic shall be present unless otherwise specified.

- Stripes and plaids – They must match at all seams unless otherwise specified.
- Plackets – No puckering at seams, especially at bottom or base is allowed.
- Hems and edge finishing – They must be even with no raw edges.
- Painted hardware – There must be less than 0.06% lead by weight.
- Needle cutting–Correct needle size and type for fabric must be used. Ball point needles should be used with knits to prevent needle cutting.
- Spandex–No exposed or cut strands of spandex yarns are allowed.
- Drawstrings – They must be secured/finished at both ends.
- Children's garments:There must be no hood or neck drawstrings on garments size 2T-12. Waist/Bottom drawstring on age grades 2T-16 may not exceed 3 inches in length outside the drawstring channel when garment is expanded to its fullest width. No toggles, knots, or attachments at the free end are allowed. Drawstrings must be bartacked at centre back so string cannot be pulled out.

III.   Care labelling:

These items must meet all requirements of the care labelling rule. Provision of practical care instructions is required.

- All care labels must remain legible and securely attached through the laundering cycle and for the life of the garment.
- They will be determined by the testing lab upon completion of all colourways of actual production sample.
- All trims and hardware must be able to withstand selected care method.
- Labels must be bi-lingual (English/Spanish)

IV.   Product labelling: The garment must be labelled with:

- Brand name
- Fibre content
- Garment size
- Country of origin
- RN# or WPL#; or the name and address of the manufacturer or distributor.

V.   Size and fit

- Technical Designer shall verify and approve as per desired specification.

VI. *Performance standards*

A. Fabric performance properties –Appropriate fabric performance standards as mentioned in section 2.3.

B. Seam performance properties

Test	Method	Requirement
Woven seams	ASTM D 1683	
Fabric < 3.5 oz/sq yd		15 lbs./in
Fabric ≥ 3.5 oz/sq yd		20 lbs./in
Knitand non-woven seams	ASTM D 1683 (Modified)	
Fabric < 3.5 oz/sq yd		30% Elongationor 6 lbs tension
Fabric ≥ 3.5 oz/sq yd		50% Elongationor 7 lbs tension
Seam slippage	ASTM D434	15 lbs./in

C. Strength at stress points and of applied decorations

Test	Method	Requirement
Reinforced stress points	ASTM D 1683 (Modified)	20 lbs
Applied decorations	ASTM D 1683 (Modified)	
(i) for 18-month age and up		15 lbs min @ 10 sec
(ii) for under 18-month age		10 lbs. min @ 10 sec
Woven pocket strength	ASTM D 1683 (Modified)	10 lbs
(i) Under 8 oz/sq yd	Non-functional	15 lbs
	Functional	
(ii) Over 8 oz/sq yd	Non-functional	10 lbs
	Functional	20 lbs
Knit pocket strength	ASTM D 1683 (Modified)	5 lbs
	Non-functional	10 lbs
	Functional	

D. Button and snap strength properties

Test	Method	Requirement
Anchored strength	ASTM D	
(i) for 18-monthage and up	1683(Modified)	15 lbs min @10 sec
(ii) for under 18-monthage		10 lbs min @10 sec
Button impact resistance	MTL S-1001	5.5 in oz (no failure noted)
Snap/desnap	ASTM D 4846	Opening 2–5 lbs Closing 2–10 lbs

E. Zipper strength properties

Test	Method	Requirement
Seams	ASTM D 2061	20 lbs
Top/bottom stop (open and closed position)	ASTM D 2061 (Modified)	15 lbs

F. Elastic must retain strength and resiliency following exposure to UV light, sun tan lotions and oils, salt and chlorinated pool water.

G. Appearance

Appearance tests include the change in garment appearance after repeated washings. This encompasses but is not limited to:

• Corrosion resistance (metal hardware)	Must display no corrosion
• Lead content of painted hardware	< 0.06%
• Ironing (If recommended)	No colour change
	Good appearance
	Good shape retention.

- Good dimensional stability.
- Retention of fabric stretch and recovery properties
- No chipping, discolouration, or rusting of hardware.
- No fraying or ravelling of buttonholes, trims, or hems.
- No twisting of seams.
- No seam openings.
- No objectionable frosting, pilling, and snagging.
- No deterioration of elasticised areas.
- No deterioration of buttons, snaps, or zippers.
- Torque shall be less than 5% of side seam length.
- Differential shrinkage difference shall not be more than 2% between shell and lining.

References

1. Shishoo R L (1995), 'Importance of mechanical and physical properties of fabrics in the clothing manufacturing process', *Int J Clothing Sci & Tech*, 7, 35-42.
2. Bureau Veritas consumer products services (2002), 'Open The Door to Quality', France, 21-46.
3. Kelley Nancy (2008), ' You don't always get what you pay for: evaluating quality of apparel'. Available from: www.textilefabric.com [Accessed 12 February 2009].
4. Kmart vendor standards (2000), Minimum performance standards for fabrics, Kmart quality assurance department, USA.
5. Kmart vendor standards (2000), Garment standards, Kmart performance standards, USA.
6. 16 CFR part1610 Standard for the flammability of clothing textiles.
7. ASTM D1230 Standard test method for flammability of apparel textiles.
8. 16 CFR part 1303 Ban of lead containing paint and certain consumer products bearing lead containing paint.
9. ASTM D 1683 Standard test method for failure in sewn seams of woven apparel fabrics.
10. MTL S-1001, Button impact resistance, Merchandise Testing Laboratory in-house test method.
11. ASTM D 4846 Standard test method for resistance to unsnapping of snap fasteners.
12. ASTM D 2061(Modified) Standard test methods for strength tests for zippers.
13. AATCC 143 Appearance of apparel and other textile end products after repeated home laundering.
14. ASTM D434 Standard test method for resistance to slippage of yarns in woven fabrics using a standard seam.
15. AATCC 22 Water repellency: spray test.
16. AATCC 35 Water resistance: rain test.
17. 16 CFR parts 1615 and 1616 Standard for the flammability of children's sleepwear: sizes 0 through 6X; Standard for the flammability of children's sleepwear: sizes 7 through 14.
18. ASTM D 4770 Standard test method for appearance and integrity of highloft batting after refurbishing.
19. ASTM D 4524 Standard test method for composition of plumage.
20. ASTM D 4522 Standard performance specification for feather and down fillings for textile products.

CHAPTER 3

Benchmarking of quality in apparel

Abstract

Quality of apparel is benchmarked as per the product performance in actual use. Requirement standard is dependent on the country of export. The chapter first discusses on the general testing requirements and related tolerances in fibre, and care labelling and flammability. The chapter then deals with the durability, performance in colour fastness, visual appearance, and various functional properties of apparel for different export markets.

Keywords: fibre, apparel, labelling, skewness, dimensional stability, AATCC

Chapter contains (Section headings)

3.1 General requirements
3.2 Flammability requirement of apparel for export
3.3 Performance in colour fastness of apparel
3.4 Characterisation of apparel durability
3.5 Performance and functional properties of apparel
References

3.1 General requirements

Major markets in the apparel trade can be broadly classified under two groups, i.e., US and non-US-based cluster. However, requirements may or may not vary accordingly. General testing requirements of fibre and care labelling in the apparel export market[1] is shown in Table 3.1. In case of fibre labelling, no tolerances exist for products made wholly of one fibre. Such product should be labelled as "100%" or "All". But there is a ±3% tolerance, by weight, for products composed of more than one fibre. In care labelling, dimensional stability[2] in both washing and drycleaning is important. The shrinkage requirement limit varies from –3% to –4% and +3% for warp as well as weft in woven goods, whereas permitted variation in knitted goods is ±5% both

Table 3.1 General testing requirements of fibre and care labelling for major apparel markets

Test parameters		US	Canada	UK	Europe	Australia	Japan
Fibre composition	Single fibre content	No tolerance	No tolerance	No tolerance	No tolerance	No tolerance	Comply with Japan fibre-labelling rule
	Blend	±3%	±5%	±3%	±3%	±3%	Comply with Japan fibre-labelling rule
Care labelling							
Dimensional stability		After 3/5 wash	After 3/5 wash				
i. Washing	Woven (warp & weft)	− 3.5 %/ + 3%	− 3.5 %/ +3%	− 4%/+3%	− 4%/+3%	− 3.5%/ +3%	±3%
	Knitted (length & width)	± 5%	± 5%	± 5%	± 5 %	± 5 %	± 5%
ii. Drycleaning	Woven warp & weft)	± 2.5%	± 2.5 %	± 2.5 %	± 2.5 %	± 2.5 %	± 2%

(Contd.)

Test parameters		US	Canada	UK	Europe	Australia	Japan
	Knitted (length & width)	± 3%	± 3 %	± 3 %	± 3 %	± 2.5 %	± 3%
Spirality		5%	5%	5%	5%	5%	5%
Colour fastness:							
i. Washing	Colour change	4	4	4	4	4	4
	Staining	3	3	3–4	3–4	3–4	3–4
ii. Drycleaning	Colour change	CC4 CS4	4	4	4	4	CC4 CS4
iii. Chlorine bleach	Colour change	4	4	4	4	4	4
iv. Non-chlorine bleach	Colour change	4	–	–	–	–	–
Garment appearance	Retention after washing or drycleaning	No noticeable shape distortion nor colour change	No noticeable shape distortion nor colour change	No noticeable shape distortion nor colour change	No noticeable shape distortion nor colour change	No noticeable shape distortion nor colour change	No noticeable shape distortion nor colour change

in course and wales direction. Similarly, in drycleaning, tolerance in woven products varies from ±2% to ±2.5% for warp and weft and ±2.5% to ±3% for knits in course and wales direction. Spirality[3] is nothing but twisting of fabric in a garment after laundering. This term is also used interchangeably as torque or skewness. The origin of spirality is from fibre, yarn and fabric construction. Molecules in the fibre tend to go back to the way it was grown or made when distorted. This is called "memory effect"[4] and is predominant in the occurrence of spirality. This phenomenon is crucial in the apparel market for knitted goods. In general, requirement standard of spirality is 5% irrespective of any market. Apart from dimensional stability, colour fastness to washing and drycleaning as well as chlorine and non-chlorine bleach are important parameters to verify or establish care label. Non-chlorine bleach test[5] requirement as rating 4 is only employed for US care labelling recommendation. For all the apparel markets, washing and drycleaning rating vary from 3 to 4 in general. Garment appearance[6] after washing and drycleaning is visually judged for shape distortion and colour change in garment trade. Wash and wear test is applicable on the durable press garment[7] which is accepted as 3.5 rating for apparels destined to any major market.

3.2   Flammability requirement of apparel for export

General testing requirements of flammability for major apparel markets[1] are mentioned in Table 3.2. Wearing apparel before export to USA must meet the requirements as mandated by the United States Consumer Product Safety Commission, i.e., 16 CFR part 1610[8] Under this Code of Federal Regulation, fabric must meet Class 1—normal flammability requirement standard with no unusual burning characteristics. In case of Canada and Sweden, time of flame spread in the apparel varies from 3.5 sec to 5 sec depending on the fabric characteristics. The code of regulations established for children's sleepwear is more stringent than general wearing apparel. In case of USA zone, the general requirements as mandated by the United States Consumer Product Safety Commission (CPSC), 16 CFR parts 1615/1616[9] are to be satisfied. However, performance requirement for the export to Australia is regulated by a different standard[10] In case of nightwear, no special regulation is there for US-based market except 16 CFR 1615/1616. But for UK market, there exists a regulation[11] The earlier Nightwear (Safety) Regulations 1985 imposed requirements relating to the flammability performance of nightwear (which includes any babies' garments), its testing and labelling. As per the latest regulation, babies' garments are not required to comply with the flammability performance requirements but must be labelled so as to indicate whether or not they are capable of complying with those requirements.

Table 3.2 General testing requirements of flammability for major apparel markets

Flammability	US	Canada	UK	Europe	Australia	Japan
Wearing apparel	CFR Part 1610 class 1	Time of flame spread : Plain fabric – more than 3.5 sec Raised surface – more than 4 sec	–	Sweden: Time of flame spread : more than 5 sec.	–	–
Children sleepwear	CFR Part 1615 CFR Part 1616	Comply with children's sleepwear regulation	–	–	Performance requirement based on AS1249-1990	–
Nightwear	–	–	Comply with nightwear safety regulation 1985	–	–	–

3.3   Performance in colour fastness of apparel

Major source of customer complaint in apparel market generates from colour fastness of textile products. The fastness of a colour is dependent on the type of dye, depth of shade, colour, and process parameters in dyeing. Dyes react differently when in contact with different agents, for instance, dyes, which may be fast to drycleaning, may not be fast to rubbing or water or perspiration. Keeping such events in view, evaluation of fastness of colours on dyed and printed textile products are mandatory requirement in export. There are number of factors that the coloured items may encounter during its lifetime which can cause the colour either to fade or to bleed onto an adjacent white or light coloured item. For all practical purposes, the effect of light, washing, drycleaning, water, perspiration, rubbing/crocking, sea water and chlorinated water cannot be overlooked. In case of colour transfer from the surface of coloured textile material to other surfaces by rubbing/crocking, the desired dry and wet staining are 4 and 3, respectively for apparel markets in the globe as depicted in Table 3.3. But for perspiration and water fastness, rating level varies from 3 to 3–4 in colour staining to the multifibre strips with regard to US and non-US-based markets, respectively. However, rating of colour change in major apparel markets remain at 4 in perspiration and water fastness. Colour fastness to chlorinated water and sea water are important for swimwear and beachwear. In both the cases, rating is 4 as far as the colour change is concerned for both US- and non-US-based market. But requirement of staining on multifibre is slightly relaxed in non-US cluster, i.e., 3–4 with reference to 3 as desired in US-based market. Colour fastness to light[12] is an important parameter to decide the quality of garment when exposed to different forms of light. Accepted rating of lining/underwear is normally 4 in US-based market but a more relaxed rating, i.e., 3 is acceptable to the non-US segment. While outerwear is normally accepted at the rating 4 irrespective of any segment of major apparel market, requirement rating differs in swimwear category in which grade 4 and 5 are considered as acceptable to the US-based and non-US-based market, respectively.

3.4   Characterisation of apparel durability

The performance of any kind of apparel can be characterised through various physical parameters. The reasons of performing such tests are many but in the apparel testing it is to obtain some indication of probable performance in use. Interaction of fibre, yarn, and fabric properties is important in such evaluation. In the apparel industry, normally performance tests are based on tensile, tear, bursting, and seam properties. The tear strength of a fabric depends on various factors. Controversial issues often heard on this property during selection of fabrics in apparel export. Some of the important points which are important but not limited to the following facts[13] such as: higher the value of single thread strength, higher is the tear strength; plied yarn gives higher tear strength than that of single yarn; twill weave gives higher tear strength than the plain weave, as twill weave has higher float which gives more grouping of the threads; and high set fabric preclude thread movement, therefore, the assistance by thread grouping is greatly reduced. In garment industry, stitching of different areas of a product is a key character to determine the quality. The efficiency of which depends on strength, elasticity, durability, security, and appearance of the constructed seam balanced with the properties of the material to be joined. Seam strength/slippage has been considered as extensively used parameter in the apparel trade for acceptance testing of a product manufactured under a particular international brand. Tensile and bursting strength properties are more frequently used parameters to characterize an apparel to predict the useful life in wearing. Due to the nature of test, the former is used for woven and the later one is applicable for knitted goods. General requirements of performance tests in major apparel markets[1] are reported in Table 3.4. It is apparent from the Table that though unit of expression is different in US- and non-US-based market for the merchandise mentioned, but almost same requirement level persists if values are converted and compared on the similar unit. This is applicable for tensile, tear, seam strength, and seam slippage properties. But in case of bursting strength, benchmark is slightly higher in US-based apparel market on similar converted unit of measurement. Interestingly, requirement levels vary depending on the product category irrespective of export to US- or non-US-based market.

3.5 Performance and functional properties of apparel

Apart from different strength properties, performance of apparel is also assessed from the point of view of actual service and some optional functional properties. Thus, general testing requirements of abrasion, pilling, water repellency, and resistance behaviours are expressed in Table 3.5 with regard to major apparel markets. It is well known that abrasion is just one aspect of wear and is the rubbing away of the component fibres and yarns of the fabric. The evidence concerning the various factors that influence the abrasion resistance is contradictory. But the main factors, which have been found to affect abrasion, are fibre type, fibre properties, yarn twist, and fabric structure.

Table 3.3 General testing requirements of colour fastness for major apparel markets

Colour fastness tests	Colour change / staining	US	Canada	UK	Europe	Australia	Japan
Rubbing/ Crocking	Dry staining	4	4	4	4	4	4
	Wet staining	3	3	3	3	3	3
Perspiration	Colour change	4	4	4	4	4	4
	Colour staining	3	3	3–4	3–4	3–4	3–4
Water	Colour change	4	4	4	4	4	4
	Colour staining	3	3	3–4	3–4	3–4	3–4
Chlorinated water	Colour change	4	4	4	4	4	4
	Colour staining	–	–	–	–	–	–
Seawater	Colour change	4	4	4	4	4	4
	Colour staining	3	3	3–4	3–4	3–4	3–4
Light	10-h exposure (lining/ underwear)	4	4	3	3	3	3
	20-h exposure (outerwear)	4	4	4	4	4	4
	40-hr exposure (swimwear)	4	4	5	5	5	5

Among different types of abrasion, plain or flat type is commonly used to determine the service performance. US-based market is not very specific to this and that is why no separate requirements are specified for different apparel products as mentioned in Table 3.5. But the situation is quite different in non-US-based market, wherein different requirement levels are specified for each category of product.

Pilling is a condition that arises in wear due to the formation of little "pills" of entangled fibre clinging to the fabric surface giving it an unsightly appearance[14] The amount of pilling that appears on a specific fabric in actual wear will vary with the individual wearer and the general conditions of use. Consequently, garments made from the same fabric will show a wide range of pilling after wear which is much greater than that shown by replicate fabric specimens subjected to controlled laboratory tests. Both US and two of the non-US-based market segments such as Australia and Japan are inclined to Random Tumble Pilling Test[15] wherein fabric specimens are subjected to a random rubbing motion produced by tumbling specimens in a cylindrical test chamber lined with a mildly abrasive material. In order to form pills that resemble those produced in actual wear in appearance and structure, small amounts of grey cotton lint are added to each test chamber with the specimens. A subjective performance rating of 3–4 in comparison with a set of photographic standards is acceptable in such kind of testing. Apparel market in non-US-based cluster has adopted ICI Pilling Test[16] as their base for evaluation of pilling phenomenon. In this test method, fabrics samples mounted on a polyurethane tube are tumbled together in a cork-lined box and evaluated subjectively by comparing it with photographic standards of a written scale of severity. A rating of 3–4 is acceptable in non-US-based market for this kind of evaluation.

Performance of certain garments for special application is evaluated by water repellency, i.e., spray test and water resistance, i.e., rain test. Spray test is especially suitable for measuring the water-repellent efficacy of finishes applied to apparels. A spray rating chart is used to determine the grade. It is well established that for a given AATCC rating there is equivalent ISO rating. For instance, AATCC rating 90 is equivalent to ISO 4 to measure the wetting behaviour of the sample under test. According to this norm, both US- and non-US-based apparel market requirements are same in original state. While after wash requirement is AATCC rating 70 for US-based market, no such standard exists for non-US-based arena except Japan. Rain test is only applicable for US-based market to predict the rain penetration resistance of fabrics. Under chapter 62 of harmonised tariff schedule of the United States, "water resistant" means that such

TABLE 3.4   General testing requirements of performance tests for major apparel markets

Test parameters	Items	US	Canada	UK	Europe	Australia	JAPAN
Tensile strength	Blouse	25 lb	120 newton	12kg	12 kg	12 kg	12 kg
	Shirt/dress/skirt/ pyjamas/lining	30 lb	150 newton	15 kg	15 kg	15 kg	15 kg
	Jacket/coat/ vest	37lb	170 newton	17 kg	17 kg	17 kg	17 kg
	Pocketing	50lb	230 newton	23 kg	23 kg	23 kg	23 kg
	Dungarees/ overall/trousers/ shorts/jeans	50lb	230 newton	23 kg	23 kg	23 kg	23 kg
Tearing strength	Blouse	1.5lb	7 newton	700 g	700 g	700 g	700 g
	Pyjamas/lining	1.8lb	8 newton	800 g	800 g	800 g	800 g
	Shirt/dress/skirt/ jacket/coat/vest	2.0lb	10 newton	1000 g	1000 g	1000 g	1000 g
	Pocketing	2.5lb	12 newton	1200 g	1200 g	1200 g	1200 g
	Dungarees/ overall/trousers/ shorts/jeans	2.5lb	12 newton	1200 g	1200 g	1200 g	1200 g

(Contd.)

Test parameters	Items	US	Canada	UK	Europe	Australia	JAPAN
Bursting strength	Diaphragm	50 lbs/inch^2	50 lbs/inch^2	2.8 kg/cm^2	2.8 kg/cm^2	2.8 kg/cm^2	2.8 Kg/cm^2
Seam properties (Strength & slippage)	Blouse/ shirt/ dress/skirt	22&15lb	22&15lb	10&7kg	10&7kg	10&7kg	10&7kg
	Pyjamas/lining	25&18lb	25&18lb	12&8kg	12&8kg	12&8kg	12&8kg
	Jacket/coat/ vest	30&22lb	30&22lb	15&10kg	15&10kg	15&10kg	15&10kg
	Pocketing	30&22lb	30&22lb	15&10kg	15&10kg	15&10kg	15&10kg
	Dungarees/ overall/trousers/ shorts/jeans	37&25lb	37&25lb	17&12kg	17&12kg	17&12kg	17&12kg

Table 3.5   General testing requirements of other performance tests for major apparel markets

Test		US	Canada	UK	Europe	Australia	Japan
Abrasion resistance	Blouse/ shirt/dress/skirt	–	–	10 000 rubs	10,000 rubs	10 000 rubs	10 000 rubs
	Pyjamas/lining	–	–	10 000 rubs	10 000 rubs	10 000 rubs	10 000 rubs
	Jacket/coat	–	–	20 000 rubs	20 000 rubs	20 000 rubs	20 000 rubs
	Pocketing: trousers	–	–	30 000 rubs	30 000 rubs	30 000 rubs	30 000 rubs
	Others	–	–	20,000 rubs	20 000 rubs	20 000 rubs	20 000 rubs
	Dungarees/overall/shorts/ jeans/trousers						
	Ladies	–	–	25 000 rubs	25 000 rubs	25 000 rubs	25 000 rubs
	Men's/children's	–	–				
				30 000 rubs	30 000 rubs	30 000 rubs	30 000 rubs
Pilling resistance	Random tumble pilling	3–4	3–4	–	–	3–4	3–4
	ICI pilling	–	–	3–4	3–4	3–4	3–4
Water repellency	Spray test: original	90	90	4	4	4	90
	after wash	70	70				70
	Rain test	Max 1.0 g water absorption	–	–	–	–	–

garments must have a water resistance so that, under a head pressure of 600 mm, not more than 1.0 g of water penetrates after 2 min when tested in accordance with AATCC test method 35-1994.

References

1. Intertek Testing Services Hong Kong Ltd. (1998), 'International apparel', Buyers' quality guide, fourth edition, Hong Kong, 79-82.
2. Anand S C, Brown K S M. Higgins L G, Holmes D A, Hall M E and Conrad D (2002), 'Effect of laundering on the dimensional stability and distortion of knitted fabrics', *AUTEX Res J*, 2, 85-100.
3. Das Subrata (2008), 'Studies on causes and remedial measures of spirality in knitted fabrics', *Asian Text J*, 17, 45-48.
4. Yan Liu, Aggie Chung, JinLian Hu, and Jing L V (2007), 'Shape memory behavior of SMPU knitted fabric', *J Zhejiang Univ - Science A*, 8, 830-834.
5. AATCC 172 Colorfastness to powdered non-chlorine bleach in home laundering.
6. Fan J, Hui C L P, Lu D, and MacAlpine J M K (1999), 'Towards the objective evaluation of garment appearance', *Int J of Clothing Sci and Tech*, 11, 151-160.
7. AATCC Test Method 124 Appearance of fabrics after repeated home laundering.
8. 16 CFR part1610 Standard for the flammability of clothing textiles.
9. 16 CFR parts 1615 and 1616 Standard for the flammability of children's sleepwear: sizes 0 through 6X; Standard for the flammability of children's sleepwear: sizes 7 through 14.
10. AS 1249 Children's nightclothes having reduced fire hazard (Foreign standard).
11. The Nightwear (Safety) (Amendment) Regulations 1987, ISBN 0110785134
12. AATCC Test method 16, option 3, Colorfastness to light.
13. Booth, J.E. *Principles of Textile Testing* (1996), Third edition, CBS Publishers & Distributors Pvt. Ltd.., India, 436.
14. Saville, B P (2002) *Physical Testing of Textiles*, The Textile Institute, Woodhead Publishing Limited, Cambridge, England, 186.
15. ASTM D 3512 Standard test method for pilling resistance and other related surface changes of textile fabrics: Random tumble pilling tester.
16. BS EN ISO 12945-1 - Determination of fabric propensity to surface fuzzing and to pilling: Pilling box method.

CHAPTER 4

Importance of flammability, care label and fibre content of apparel

Abstract

The information on flammability, care label and fibre content is important in apparel sector particularly for export market to protect the interest of the consumers. This chapter first discusses the significance of essential regulations associated with flammability, care label and fibre content for different export destinations. It then describes each parameter and the guideline of their application in apparels made out of different fibres.

Keywords: flammability, care labelling, fibre content, safety regulation, federal trade commission

Chapter contains (Section headings)

4.1 Essential standards and regulations
4.2 Flammability
    4.2.1 Flammable fabrics act of US for general wearing apparel
    4.2.2 United States flammable fabrics act standard for the flammability of children's sleepwear
    4.2.3 U.K. nightwear safety regulations
4.3 Care labelling of garments
    4.3.1 Fabric care basics
    4.3.2 Basic precautions to prevent fading of colours
    4.3.3 Detergent system
    4.3.4 Ironing
    4.3.5 Drycleaning
    4.3.6 Standard care label instructions and what instructions mean
    4.3.7 Care label recommendation

4.1 Essential standards and regulations

Export of a consumer product has its own requirement to meet the standards and regulations. Apparel product is not an exception to this. Safeguards are there to take care of the concern of the consumers and to protect their interest for which they invest. Among various essential regulations if we try to emphasise few important parameters, then flammability, care labelling and fibre identification aspects need to be highlighted with reference to the major apparel markets. Regulations differ depending on the merchandise and its destination to the country of export but the basic objective is to protect the interest of the consumer.

4.2 Flammability

4.2.1 Flammable fabrics act of US for general wearing apparel

This act was developed to remove highly flammable products from commerce. As a mandatory safety regulation, Consumer Product Safety Commission (CPSC) of USA monitors all textiles used for general wearing apparel.[1]

In accordance with the Code of Federal Regulation[2] flammability of general clothing textiles is regulated. This standard measures the speed and intensity of flame and ease of ignition. Three classes of flammability have been covered in this standard for classifying textiles and warn against the use of textiles that have burning characteristics unsuitable for clothing.[3]

Class 1, normal flammability

The time of flame spread is 3.5 seconds or more for textile without nap, pile, tufting, flock or other type of raised fibre surface and more than 7 sec for textile with nap, pile, tufting, flock or other type of raised fibre surface, provided the intensity of flame is so low as not to ignite and fuse the base fabric.

Class 2, intermediate flammability

The time of flame spread is from 4 to 7 seconds for textile with nap, pile, tufting, flock or other type of raised fibre surface and the base fabric ignites and fuses. This class is although legal but is not acceptable for sale.

Class 3, rapid and intense burning

The time of flame spread is less than 3.5 seconds for textile with or without nap, pile, tufting, flock or other type of raised fibre surface. This class of fabric is dangerously flammable and not allowed to be legally sold in the United States or imported into the United States from abroad.
Few well recognised exemptions are furnished as follows:

    i. Hats, (with less than 9″ trims), gloves and footwear
    ii. Interlinings – unless garment could be worn open such as flannel lined windbreaker. If garment could be worn inside out (sweatshirt), then the backside of fabric is subject to the standard.
    iii. Plain surface fabrics weighing 2.6 ounces per square yard or more. Plain surface is defined as any fabric that does not have an intentionally raised fibre or yarn such as tufting, file or nap.
    iv. Both plain and raised surface fabrics, regardless of weight, made entirely from any of the following fibres or entirely from combination of the following fibres: acrylic, modacrylic, nylon, olefin, polyester and wool.

High risk fabrics

- Flammability test in high-risk fabric such as acetate, linen, rayon, cotton and silk is compulsory for all colourways, even the area existing on actual product is small.
- Textiles with fuzzy or napped surface
- Open weave or sheer fabrics weighing less than 2.6 oz/sqyd such as chiffon, crepe de chine, gauze, etc.

4.2.2 United States flammable fabrics act standard for the flammability of children's sleepwear

Children's sleepwear is covered under and must meet the requirements of the Code of Federal Regulation[4] that measures fabric flammability by char length of burn. These standards protect children from serious burn injuries if they are exposed to an open flame, such as a match, lighter or stove burner.

The code of regulations established for children's sleepwear (sizes 0–6 × and 7–14) is more stringent than general wearing apparel. Under federal safety rules, garments sold as children's sleepwear for babies larger than 9 months and up to size 14 must be either flame resistant or snug fitting. Some of the reputed companies prohibit the use of flame retardant finishes on any infant or children's merchandise.

Interestingly, sleepwear is defined as 'any product of wearing apparel such as robes, nightgown, sleepers and pajamas, intended to be worn for sleeping or activities related to sleeping.'[5]

Exemptions in this category are as follows:

- Diapers
- Underwear
- Snug-fitting cotton sleepwear that meets the CPSC measurements for snug-fitting
- Garments sized for infants of 9 months or younger is defined as follows:
  - A one-piece garment that does not exceed 25.75 inches (64.8 cm) in length
  - A two-piece garment with no piece exceeding 15.75 inches (40 cm) in length
  - Bearing a label stating the size of the garment expressed in terms of months of age (e.g. '0–3 months')

While above items do not need to be tested against the more stringent children's sleepwear flammability requirements, they must be tested against flammability standards for general clothing textiles, 16 CFR 1610.

4.2.3   U.K. nightwear safety regulations

Nightwear can burn rapidly, when accidentally set alight by contact with an open fire or a gas or electric fire or other heat source and causes serious injury. Various mandatory and voluntary measures have been taken to control the fire performance of the fabrics used in nightwear and to create public awareness of the dangers. The 1985 Regulations[6] impose requirements relating to the flammability performance of nightwear (which includes any babies' garments, as defined in Regulation 3(1)), its testing and labelling. The regulations came into effect on 1 march 1987 and replaced the Nightdresses (Safety) regulations 1967 and the Nightdresses (Safety) regulations (Northern Ireland) 1968. As per the regulations of 1985, nightwear types include the following types:

- *Baby garments*: Garments exclusively for babies less than 3 months and having a chest measurement not exceeding 53 cm(21″).

Table 4.1   Care symbols and sequence for US and European market

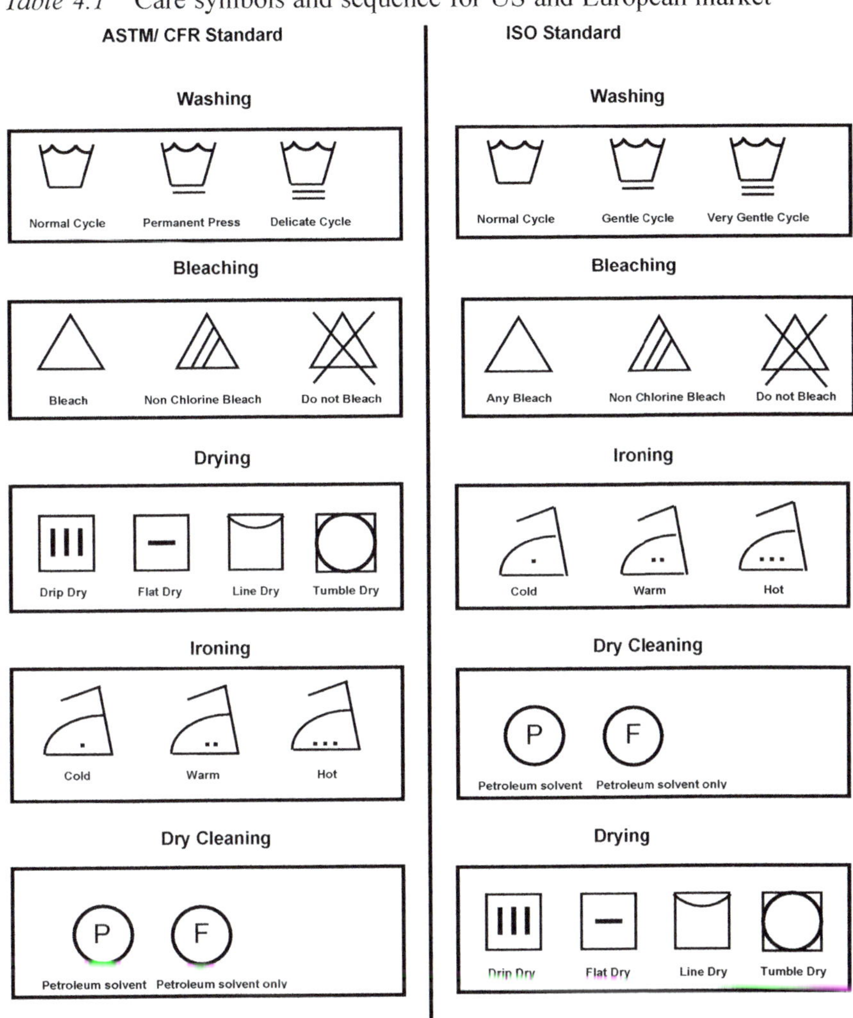

- *Children's nightwear*: Garments for children over 3 months and under 13 years of age and not exceeding any of the following maximum measurements.
  - Nightdresses: Chest measurement: 91cm (36″); Length:122 cm (48″)
  - Dressing Gowns, Bath Robes and other similar garments: Chest measurements: 97 cm (38″); sleeve measurement: 69 cm (27″)
- *Adult Nightwear*: Same type as Children's nightwear and commercially not usually tested.

Nightwear is tested to see whether or not it meets the flammability performance requirements, which are specified in clauses 3.1.1 and 3.2.1 of British Standard 5722: 1984.[7] These requirements are expressed at a rate of flame spread. The method of test to be used is Test 3 of British Standard 5438.[8] Before testing, test pieces must be washed once in accordance with the procedure specified in Clause 6.5.2 of British Standard 5651.[9] If the fabric has been treated with flame retardant chemicals to make it safer from fire, the test pieces must be washed 12 times as specified in Clause 6.5.2.7 of this British Standard. This is to ensure that the treatment is sufficiently durable. If any test piece burns to a trip thread at 300 mm (12″) above the flame point in less than 25 seconds or to a second trip thread at 600 mm (24″) above the flame point in less than 50 seconds, the test is failed.

The amended regulations[10] apply to nightwear including garments, which are commonly worn as nightwear. These Regulations amend the Nightwear (Safety) Regulations 1985. As a result, those regulations come into force on 1 September 1987 as regards the requirements relating to babies' garments. 1 March 1987 remains the date on which all other requirements of the 1985 Regulations come into force. Babies' garments are not required to comply with the flammability performance requirements but must be labelled so as to indicate whether or not they are capable of complying with those requirements.

New flammability performance requirements for children's nightwear in the UK have been introduced by the European standard BS EN 14878[11] and came into effect in November 2008. However, EN 14878 is not legislation but it is a voluntary European standard. It is worthwhile to mention that children's nightwear in the UK must comply with the Nightwear (Safety) regulations 1985, and continue to do so even after November. Some requirements of BS EN 14878 are less demanding than those of the UK regulations. However, those parts of BS EN 14878 that go beyond the requirements of the UK regulations must be carried out to truly comply with the General product safety regulations 2005 (GPSR). As per BS EN 14878, the test method for the determination of flammability is followed as per BS EN 1103[12] which is different from BS 5722: 1984. As per the new method, there are the following three classes:

- Class A (not pyjama): 520 mm trip thread not severed in less than 15 seconds and no surface flash.
- Class B (Children's pyjama): 520 mm trip thread not severed in less than 10 seconds and no surface flash.
- Class C (Babies' nightwear): not tested and no requirements.

In principle, it is recommended that the more onerous of the requirements of the UK regulations and GPSR/BS EN 14878 should be applied to children's nightwear in order to meet the statutory requirements of both the GPSR and the UK regulations.

4.3   Care labelling of garments

Apparels and textiles are soiled during normal use. Economic realities require used items must be cleaned and refurbished for reuse without substantially altering their functional and aesthetic properties[13] Principally consumers but also launderers and dry cleaners have the choice to select the correct technique to restore the attributes of the textiles. With a view to assisting consumers in getting information about clothing care, US Federal Trade Commission has promulgated care labelling rule in 1971 and amended it in 1983[14] European communities within the framework of IEC/ISO directives facilitate standard work on textile care labelling which is delineated in ISO/FDIS 3758:2003(E)[15] The rule requires manufacturers and importers of textile wearing apparel and certain piece goods to provide regular care label instructions when those products are sold. The purpose of the rule is to give the consumer accurate care information, so that the processes contained on the label will avoid any damage of the product. However, the care label should not be regarded as a quality seal. It only denotes the maximum permissible treatment without irreversible damage. Care labels must be permanently attached and remain legible for the life of the garment. Sometimes symbols may be used in conjunction with words but will not by themselves satisfy the requirements explicitly[16] Care labels, often, are deciding factors when consumers shop for clothing. While some opt for the convenience of drycleaning, others prefer the economy of buying garments, which they can wash. Some manufacturers try to reach both markets with garments that can be cleaned by either method. The care label rule allows providing more than one set of care instructions if a reasonable basis for each instruction exists. Although there is relaxation in using only temporary labels for products such as totally reversible clothing without pockets and products that may be washed, bleached, dried, ironed or dry cleaned by the harshest procedures available, but, interestingly, no care instruction is needed for some of the cases such as products sold to institutional buyers for commercial use and products which are completely washable and sold at retail for $3 or less. Care label rule requires that manufacturers and importers of textile wearing apparel have a reasonable basis and reliable evidence in support of care instruction. That is why different tests are involved while determining the care label of a garment. Recommendation of care label is associated with washing, bleaching, drying, ironing, and drycleaning for different set of conditions and evaluated for appearance as per different standards such as ISO 3758:2003, ASTM D3136:2000[17] ASTM D5489-01a[18] FTC Care labelling guide and 16 CFR part 423 (14). However, the order of representation varies depending on the standard. For instance, in export to USA, the order is represented as washing, bleaching, drying, ironing and drycleaning. But, in the case of Europe, the sequence is depicted as washing, bleaching, ironing, drycleaning and drying. Symbols that communicate care procedures may be used in addition to words, but the words must fulfil the requirement of the care label rule. A schematic representation of care symbols at the right order for ASTM/CFR (USA) and ISO (Europe) is represented in Table 4.1. Sometimes, for exports to USA, multiple care instructions are provided when one style has different colourways. This is also due to the failure of some colours in the bleaching test. Care label recommendations, in this case, depend on colour. And the judgment must be directed to prohibit the use of one standard care label for all colours. In certain exception and mainly due to commercial reason, buyers tend to use a safe care instruction, which may be applicable to all colours in the same style. However, such cases are obviously considered as violation of the FTC rules and regulations for care labelling.[19]

4.3.1 Fabric care basics

In today's fashion world, dark clothes dominate the platform – especially the basics. Solid black, brown and navy are wardrobe essentials. But how do we negotiate with fabric fading? What are the laundering precautions to keep garments looking new? Here are few probable answers.

4.3.1.1 Causes of colour loss

Quality of dyes: Dye performance is determined by nature of dye and application method. Different types of textile substrates do play a major role in dye uptake mechanism. A mistake or mismatch in any of these can cause garments to lose/bleed colour. Complaint of the consumer is the end result.
Temperature: There is every chance of faster colour loss/bleeding if a garment is washed in water that is warmer than recommended.
Drying condition: It's a mistake to over-dry dark colour clothes or dry them at a too-high temperature.

4.3.2 Basic precautions to prevent fading of colours

Referring care labels: Before treating any garment, always first refer to the instructions on the care label. Symbols on garment labels are there for a reason and prolong the life of clothes.

Judicious sorting: Success in laundry starts with sorting items by fabric type, colour and temperature of wash. Laundry items can be sorted into five main groups:

- whites–everything white, like underwear, t-shirts, handkerchiefs, etc.
- lights–including striped whites, off-whites and pastels

- darks–everything dark, like blacks, blues, and browns
- brights–reds, yellows, oranges, and fluorescents
- delicates–fine linens, lingerie, and some synthetic fabrics

Also, heavily soiled items are advisable to be separated out from lightly soiled garments, and it is better to shake out loose dirt.

Turn inside-out: Turning garments inside-out before washing and drying reduces abrasion (a major culprit in fading) and prevents the dulling effect due to pilling.

Use the right products: It is necessary to follow product instructions as carefully as one follows care label instructions. The use of too much or too little detergent can cause dulling.

Use right loading: Overloading the washer or dryer will not allow clothes to move freely, allowing for detergent deposits or poor rinsing/residue removal. Machines operate better when clothes are evenly distributed and balanced.

Water temperature: Washing and rinsing in cold water protect darks better than warm or hot water.

Careful drying: Over-drying (especially with lighter, lint-producing garments) can unnecessarily wear out dark colours. It is always better to remove garments slightly damp and leave them inside-out to dry.

Avoid sunlight: Sunlight exposure, in drying or in storage, may destroy colour. Therefore, it is best to dry/store in shed.

4.3.3   Detergent system

Detergent systems in commercial laundering were traditionally been of five components, i.e. (i) alkaline agent: to raise pH for cleaning, (ii) detergent: for actual cleaning, (iii) chlorine bleach: to destroy various coloured stains, (iv) souring agent: to provide acidity and to lower ph back to acceptable level and (v) softener: to improve hand/feel. In the smaller capacity laundering, the alkaline agent and detergent are combined into one and/or the souring agent and the softener can be combined into one. This is leading to either four or three product system.[20]

In recent days, low alkalinity laundering systems have been used by utilizing enzyme and more high-tech cleaning chemistry to substitute for old-fashioned high alkalinity system. High-tech cleaning chemistry involves affinity (hydrophilic–lypophilic Balance, i.e., HLB value of a detergent to handle oil, fat, and grease), polarity (ionic nature of detergent, i.e., cationic, anionic, non-ionic, and amphoteric), surface tension (necessary to reduce surface tension to make water wetter), emulsification (to be able to suspend and disperse oil) and pH (power of hydrogen to increase the effectiveness of the cleaning process). Interestingly, the need of souring has been also been eliminated in simplified three product system of formulated detergent, bleach and softener. Most of the commercial synthetic detergents are the derivatives of petroleum oil fractions.

Experts used to agree that the best cleaning results were achieved by washing items in the hottest water temperature that the fabric will allow. But water temperature —whether it's hot, warm or cold —affects the performance of laundry products on soil removal, fabric wrinkling and shrinking, colour bleeding and fading and overall durability of fabric finishes. Laundry code symbols on clothing tags play an important role in the selection of right temperature in washing.

Temperature	112°F–145°F	87°F–112°F	65°F–86°F
Suitable for:	• Whites • Heavy soiled fast coloured fabrics • Towels/washcloths • Oily, greasy dirt	• Dark colours • Colourfast brights • Permanentpress	• Delicates • Fabrics with dyes that may run or bleed • Lightly soiled items

Important points to remember

It is advisable to wash reds or new, coloured garments separately the first few times. These items can bleed and stain other laundry if not careful. To test an item for tentative colourfastness, it can be dampened with water in a discreet spot and blot with an old white cloth. If colour transfers to the white, the item will bleed. To be sure, an old white handkerchief or sock can be added with the possible offenders until it comes out clear, then one can wash those items with other like colours without fear of bleeding.

All clothes need to be checked for stains and those that require pretreatment or soaking can be sorted out. Liquid detergent is then transferred to a spare container to pre-treat clothes.

All pockets and pant cuffs are to be carefully checked for things those need not to be washed, and even the inside of the machine can be inspected for the same.

Zippers, buttons, snaps and buckles also are to be observed and secured to prevent snagging. Shirt cuffs are to be unrolled, drawstrings shall be tied and un-removable shoulder pads are to be secured.

Mesh bags are used to separate washable delicates from rougher fabrics or to designate items, which cannot be transferred to the dryer.

Lint generators and lint magnets are not to be mixed. Some lint generators include towels, sweatshirts and flannel. Lint magnets include corduroy, velvets and permanent-press clothes. When in doubt, it is better to turn the lint-magnet items inside out during sorting.

Fabrics, linings or insulations could shrink or be otherwise damaged from using too hot a temperature or from leaving the garments in the dryer too long. Apparels made from fibres such as acrylic, nylon, polyester and polyolefin tend to dry quickly and thus to be watched carefully. Polyolefin can actually melt if the dryer temperature gets too high.

4.3.4 Ironing

Ironing, pressing and finishing are the terms, which are often used interchangeably. The basic aim of these different activities is to remove the 'unwanted' crease and impart 'wanted' creases. Ironing involves relative movement or friction between two flat surfaces, while pressing involves compression between two surfaces with no relative motion between them. Finishing removes the 'unwanted' crease by stretching, but cannot impart a 'wanted' crease. Both ironing and pressing can remove or impart a crease. Ironing is one job most users love to hate. It can take a lot of time and energy, neither of which are in abundance in today's hectic households. Even though wrinkle-free fabrics are more readily available in market, ironing is never going to disappear. So if one wants his clothes to have that finished look but donot have the ironing know-how or time, following basic ironing warnings may be surely of help to him. Ironing actually begins in the washer and dryer. Synthetics made of nylon, polyester and acrylic and washable woolens are to be washed in hot or warm water using a permanent-press cycle to remove and reduce wrinkles. Bright colours and lightly soiled fabrics are to be washed in cold water to minimise washer wrinkling and to save hot water. The addition of an appropriate wrinkle control agent in the rinse cycle also helps reduce wrinkles. One can shake out items taken from the washer before placing them in the dryer to prevent them from balling up and wrinkling. The dryer shall be correctly loaded to prevent improper tumbling, causing clothes to dry slower and wrinkle. Over drying of clothes not only causes shrinkage but also increases static cling. Excessive heat can also set wrinkles.

Hang up or fold clothes immediately after removing them from the dryer is required. If one leaves them in a heap, they will wrinkle.

Rule of ironing of garment is interesting and the movement of iron over the garment follows certain rules given as follows:

- From right side of the ironing table to the left – requires steam and no suction, and
- From the left side of the ironing table to the right – no steam, dry iron and suction.

Vacuum and steam should never be applied together. A synergistic rhythm of three applications such as movement of the iron, application of steam and application of vacuum shall generate effective result in ironing. In a practical erroneous situation, i.e. applying steam while the vacuum is on, may not result in any quality issue of garment but it affects energy bill and incorrect motion slows down productivity.

Undoubtedly, some fabrics,such as cotton or silk, will beg for an iron. Laundry care symbols on the label will indicate on what temperature to set the iron. One can also consult fabric guide information for selection of iron temperature. Suggested guidelines can also be followed:

Fabric	Instruction
Acetate, acrylic	Cool iron
Cotton, linen, ramie	Use steam with medium and high heat
Nylon	Low heat
Polyester	Low or medium heat
Rayon	Iron inside out on low heat
Silk	Iron inside out on low heat
Wool, mohair, cashmere, camel, alpaca	Use steam and medium heat

Garments that are dirty or stained should not be ironed because heat can set the stain.

For better result, one can iron items that need lower iron temperatures first, and then end with those requiring higher temperatures. Ironing of clothes, especially those made of cotton, rayon and silk, is better executed while they are still damp by removing them from the dryer before they are completely dry. If that's not convenient, one can dampen dried clothes with a steam iron or sprinkle with warm water to allow the moisture to permeate the fabric. Keeping a damp sponge or spray bottle is always handy when ironing. If one creates a crease, dampening it and then re-ironing the area are advised.

Ironing of fabric on the wrong side or use a pressing cloth on the right side to avoid shine marks is always beneficial. Newly ironed items should be hanged immediately because they tend to wrinkle again quickly.

4.3.5   Drycleaning

Drycleaning is the use of solvents to remove soil and stains from fabric. It is called 'drycleaning' because the solvents contain little or no water and do not penetrate the fibres as water does. Drycleaning solvent is not harmful to most fabrics compared to water, cause less shrinkage, colour fading and other problems that can occur during the cleaning process. And, overall, solvents provide better cleaning potential than water. Drycleaning is the only safe method for cleaning many types of garments. It helps in protecting the expected life of a garment.[21]

Historically, Stoddard solvent, Carbon Tetrachloride and Valelene 113/ Freon 113 were used as the drycleaning solvents. Modern reagents include Perchloroethylene High flash point hydrocarbons DF-2000 (140°F flash point), Modified hydrocarbons blends (Pure Dry), Glycol Ethers (Dipropylene glycol tertiary-butyl ether) (Rynex), Cyclic Silicone (Decamethylcyclopentasiloxane) (GreenEarth) and Supercritical Carbon Dioxide. Perchloroethylene is undoubtedly the most commonly used solvent with unmatched cleaning performance though it is considered as a persistent and bio-accumulative chemical that is toxic to environment. Detergents are utilised in drycleaning solvents for enhanced cleaning capability. Sometimes, sizing chemical is added to restore garment shape, body and texture.

Natural fibres such as wools and silks will shrink and perhaps lose their colour when washed in water but will dry clean beautifully. Cottons and linens, unless they are preshrunk in manufacture, will also shrink in home laundering. Drycleaning is particularly effective in removing greasy, oily stains from synthetic fibres, which have an affinity for oils.

But the professional dry cleaner provides more than just drycleaning. This service also includes professional removal of problem stains that will not come out with simple drycleaning. It also includes professional pressing, careful packaging and inspections at every step along the way to make sure that all stains have been attended to and the item has been properly pressed and finished.

Great technological advances have been made for both the improvement of natural fibres and the creation and development of synthetic fibres. Special finishes impart body, permanent press qualities, water repellency and other qualities to fabrics. Fibres are blended to obtain fabrics with the best qualities of both natural and synthetic materials. Peculiarities of various fabrics are mentioned as follows:

Many beautiful fibreslack durability and should be purchased only with this understanding. These include cashmere, camel's hair and mohair. Angora, another luxury fibre, can shrink excessively even with the most careful care in cleaning. Lightweight and loosely woven wools, gauzes and loosely knit sweaters have a tendency to snag easily or become distorted in wear and cleaning. Suede and smooth leathers have a high incidence of colour difficulties. Genuine suede and leather items require special processing to preserve their finish, feel and colour. Normal drycleaning may lead to cracking, shrinking or spotting. By adding a detergent plus conditioner in the recommended concentration to the drycleaning solution to condition it, the colour and suppleness of suede and leather can be protected and preserved so that no colour loss, bleeding, stiffening and hardening will occur in the drycleaning procedure. Even drycleaning items made of combinations of suede, leather, fur and cloth can be dry cleaned as easy as drycleaning cloth items when the detergent plus conditioner is added to the drycleaning fluid. Suede-like materials and other materials with a flocked finish may develop bare spots in wear and cleaning. The life expectancy for these garments is generally rather short. Many tailored garments contain interfacings in the collar and lapel that are fused rather than stitched to the shell fabric. In some cases, blisters and wrinkles develop when these items are dry cleaned. This is the fault of the manufacturer.

Some bonded fabrics may separate from the face fabric or lining, or there may be shrinkage, puckering, stiffening or adhesive staining. Acrylic knits are inclined to stretch when wet or when exposed to steam in finishing after drycleaning. Some dyes and pigment prints may fade in drycleaning solvents. Others are water soluble and may fade when exposed to water in spot removal.

One must also be concerned with the response of buttons, beads, sequins and other decorations and fasteners to drycleaning. Most troublesome in this respect are buttons and beads made of polystyrene, which soften or melt on exposure to drycleaning solvent. Beads and sequins may be covered with a thin coating of colour, which may come off during wear or cleaning. Beads or sequins may be merely glued on and come off during wear or in cleaning. Trim that is sewn on with a single continuous thread may all come off if the thread is broken. Belts or other items that contain cardboard stiffeners or glues will require special attention.

4.3.6   Standard care label instructions and what instructions mean

Washing process

'Machine wash': Use any type of home-type washing machine.

When no temperature is given, e.g., 'warm' or 'cold', hot water up to 150°F (66°C) can be regularly used.

Machine-wash 'hot': Set initial water temperature control at 112°F–145°F (45°C–63°C).

Machine-wash 'warm': Set initial water temperatures control between 87°F to 112°F (31°C–44°C) (hand comfortable).

Machine-wash 'cold': Set initial water temperature controls same as cold water tap up to 86°F (30°C).

'Delicate cycle' or 'gentle cycle': Machine is set for slow agitation and reduced time.

'Durable/permanent press cycle': Machine is set for cold down rinse or cold rinse before reduced spinning.

'Hand wash': Garment may be laundered through the use of water, detergent or soap and gentle hand manipulation.

'Wash separately': Alone.

'With like colours': With colours of similar hue and intensity.

Bleaching process

'Bleach when needed': Any household laundry bleach may be used when necessary.

'Only non-chlorine bleach when needed': Use non-chlorine bleach only. Chlorine bleach may not be used.

'Do not bleach': No bleach may be used.The garment is not colourfast or structurally able to withstand any bleach.

Drying process

'Tumble dry': Use machine dryer. When no temperature is given, machine drying at a hot setting may be regularly used.

Tumble dry 'medium': Set dryer at medium heat.

Tumble dry 'low': Set dryer at low heat.

'Durable press' or 'permanent press': Set dryer at permanent press setting.

'No heat': Set dryer to operate without heat.

'Remove promptly': When items are dry, remove immediately to prevent wrinkling.

'Drip dry': Hang dripping wet with or without hand shaping or smoothing.

'Line dry': After spinning in the washer, squeezing through wringer or squeezing by hand, hang damp from line or bar in or out of doors.

'Dry flat': Lay down horizontally for drying.

Ironing process

'Iron if needed': Regular iron may be needed and may be performed at any temperature with or without steam is acceptable.

'Low heat iron' or 'cool iron': Regular ironing, steam or dry may be performed at low-temperature setting (230°F, 100°C)

'Medium heat iron' 'warm iron': Regular ironing, steam or dry, may be performed at medium temperature setting (300°F, 150°C).

'High heat iron' or 'hot iron': Regular ironing, stream or dry may be performed at high-temperature setting (390°F, 200°C).

'No steam' or 'do not steam': Steam in any form may not be used, but regular dry ironing at indicated temperaturesetting is acceptable.

'Steam press' or 'steam iron': Use iron at steam setting and at the indicated temperature.

'Iron damp': Moisten articles before ironing.

'Do not iron': Item may not be smoothed or finished with an iron.

Drycleaning process

'Dry clean': Dry clean, any solvent, any cycle, any moisture and any heat. The process may include the use of petroleum, fluorocarbon or perchlorethylene; moisture addition up to 75% relative humidity; hot tumble drying up to 160°F (71°C) and restoration by steam press or steam-air finishing.

'Dry clean, reduced moisture': decreased relative humidity.

'Steam only': Employ no contact pressing when steaming.

'No steam' or 'do not steam': Do not use steam in pressing and finishing.

'Dry clean, low heat': Reduced drying temperature.

'Do not dry clean': Garment may not be commercially dry cleaned.

'Leather clean' or 'suede leather clean': Have to be cleaned only by professional cleaner who has special leather care methods.

Statements for care instructions:

Processing Stage	'Standard' or 'Special' statement for Care Instruction	Applicable areas
Washing	Turn garment inside out	Rubber print
		Reverse fleece
	Machine wash cold	colour block designs
	Gentle cycle	Knit items
		Sweaters
	Hand wash cold	'Crinkle effect' items
		Silk underwear
		Pantyhose and tights
	With like colours or wash separately	Colour staining result is below requirement in accelerated washing test (AATCC 61)

(Contd.)

Processing Stage	'Standard' or 'Special' statement for Care Instruction	Applicable areas
		Crocking result is below requirement
		Pigment print or pigment dye
		Sulphur black and indigo dye
		Dark and intense colours (Black, navy, burgundy, etc.)
	Wash once before wearing	Crocking result is below requirement at original state but results pass after one wash cycle
Drying	Line or flat dry	Garments with sequins, beads or delicate embellishments
		'Crinkle effect' items
		Wool and silk delicate sweaters
	Tumble dry low	Coated items
	With clean tennis balls	Down and feather filled items
Bleaching	Do not bleach	Non-chlorine and chlorine bleach results are below requirement
	Only non-chlorine bleach when needed	Chlorine bleach result is below requirement
		Silk, wool, and spandex items
Ironing	Do not iron	'Crinkle effect' items
		Pile or double face fabrics

(Contd.)

Processing Stage	'Standard' or 'Special' statement for Care Instruction	Applicable areas
		Synthetic swimwear
	Do not iron on print	Rubber print
	Cool iron	Coated fabrics
	Do not iron on buttons	Pearl buttons
	Do not iron on beads or sequins	Pearl or plastic beads and sequins
Drycleaning	Dry clean only	If results of home laundering is below requirement

There is no doubt that care of apparels is an important area of discussion when one considers the sea changes occurred in the fashion world. The advent of new functional finishes and application of embellishment, sequins and fancy prints have transformed the wash care issues more complicated. But the basics remain the same. Keeping the fundamentals in mind and selecting right choice of parameters underlined in the process, laundering can be executed with success. Useful life of apparel can be easily ensured if one is really interested to respect the care instruction. International brands are conscious and concerned on this aspect while arriving at an appropriate care label. But the ultimate care of the product lies with the consumer, the end user of the merchandise. Legal issues do happen due to inappropriate care labelling of garment and penalty, recall and claims are the final result.

4.3.7   Care label recommendation

4.3.7.1   Purpose

To recommend the care label in textiles which describes best practice to refurbish the product without adverse effects and warn against any part of the directions which is expected to harm the product.

4.3.7.2   Principle

Sample is subjected to different tests, i.e. washing, bleaching, drying after washing, ironing or drycleaning for different set of conditions and evaluated for appearance as per different standards such asISO 3758:1991, ASTM D3136:2000, ASTM D5489-01a, FTC Care labelling guide and 16 CFR Part 423.

4.3.7.3 Theory

Symbols should be recommended in the following order:

ASTM/CFR	ISO
Washing	Washing
Bleaching	Bleaching
Drying	Ironing
Ironing	Drycleaning
Drycleaning	Drying

In case of washing, drying and ironing symbol dots may be used to define the temperature which is specified as follows:

Washing

Dots	ASTM/CFR		ISO
.	30°C	Cold	30°C
..	40°C	Warm	40°C
...	50°C	Hot	50°C
....	60°C	Very hot	60°C
.....	70°C	Very hot	70°C
......	95°C	Very hot	95°C

In addition to the basic symbols, a bar under the symbol means that the treatment is more gentle which is described as follows:

Symbol	ISO	AATCC
	Normal cycle	Normal cycle
	Gentle Cycle	Permanent Press
	Very gentle or wash as wool	**Delicate/gentle cycle**

Drying

Tumble Dry

Dots	ISO	AATCC
.	Low	Low
..	Normal	Medium
...	-	High

Other Drying instructions:

- Line dry
- Flat dry
- Drip dry

Other drying symbols to be recommended for the US

Line dry/
hang to dry

Drip dry

Dry flat

For Europe

Sample is to be line dried/ flat dried/drip dried

Ironing

Dots	ISO	AATCC
.	110°C	110°C
..	150°C	150°C
...	200°C	200°C

4.3.7.4 Instructions

Recommended tests

Care label recommendation for the US market:

- Either washing or drycleaning instructions can be recommended.

- If a sample passes the washing instructions, there is no need to proceed for the drycleaning tests.
- Any symbol which is not reported indicates sample is safe for that particular instruction.
- The most common cleaning method for an American customer is machine wash and tumble dry. So you should not recommend the method which is more difficult unless this basic cleaning method would harm the product.

Tests to be performed:

- Colour fastness to washing
- Dimensional stability to Washing.
  Note: Report before and after ironing in case of rayon fabrics (moss crepes, georgettes, chiffons, etc.)
- Appearance after washing and ironing
- Colour fastness to Chlorine Bleach (in-house method)
  (exception on animal fibres, silk, spandex/elastic and their blends. Comment is to be given in these cases, as 'Chlorine Bleach test is not applicable because of inherent properties of the fibres in the submitted sample')
- Colour fastness to non-chlorine bleach (in-house method)
- If sample fails the wash tests proceed for following drycleaning tests:
- Colour fastness to dry clean with perchloroethylene.
- For embellished/delicate styles only, appearance (with shrinkage) after drycleaning is to be performed.
- Assess spirality in Knitted fabrics and side seam spirality in garments (woven and knits)

Tests to be performed for Care Label Recommendation for the UK and European market

(I)   General tests:

- Colour fastness to washing
- Dimensional stability to washing (Follow ISO 6330:1984)
- Appearance to washing and ironing
- Colour fastness to drycleaning
- For embellished/delicate styles only, appearance (with shrinkage) after drycleaning is to be performed.
- Assess spirality in knitted fabrics and side seam spirality in garments (woven and knits)

Table 4.2. General guideline for different fabric types.

Sl no	Fabric type with respect to fibre content	Cleaning procedure	Temperature (°C)	Cycle	Drying procedure		Ironing
					Woven	**Knitted**	
1	Cotton and cotton rich blends	Machine wash	40	Normal	Tumble dry/Line dry	Heavy knits or rib flat dry Light weight knits: single jersy: tumble dry low Interlock: line dry	Warm iron
2	Polyester and polyester-rich blends	Machine wash	40	Normal	Tumble dry low/line dry	Tumble dry low	Cool iron
3	Viscose and viscose-rich blends	Machine wash	40	Gentle	Line dry	Flat dry	Warm iron
4	Washable wool and wool-rich blends	Machine wash	40	Gentle	Line dry	Line dry/flat dry	Warm iron
5	All elastane blends	Machine wash	40	Gentle	Line dry	Flat dry	Cool iron
6	Hand wash wool Hand wash silk Chenille	Hand wash			Flat dry	Flat dry	Cool iron

(II)   For white colour samples only:

Test items are same as those listed in (I) with additional test on 'colour Fastness to Bleaching with Hypochlorite' using in-house method. For coloured goods, recommend 'Do not bleach'.

4.3.7.5   Assessment criteria

Where buyer manual is available, follow requirement from manual. Otherwise follow the following requirements.

Dimensional stability to washing

Apparels:

Woven	±3.0%
Knitted/net	±5.0%

Home furnishing items- ±5.0%
Dimensional stability to dry clean

Apparels:

Woven	±2.0%
Knitted	±3.0%

Home furnishing items: ±5%.

Colour fastness to washing

- Colour change ≥ 4

Staining on M/F USA ≥ 3.5*
        other countries ≥ 3 – 4*
Self staining ≥ 4.5 or 4 – 5
*If the staining is less than the specified then mention 'wash with like colours' or 'wash separately' in the washcare.
Appearance after wash/dry clean

- No noticeable skewing (side seam skew; woven ≤ 3% and knitted ≤ 5%)
- Slight colour change (colour change ≥ 4)
- No self staining (self staining ≥ 4.5)
- Slight pilling or fuzziness (pilling rating ≥ 4)
- No noticeable distortion in terms of shape or handle.
- No noticeable fraying of threads at seams, buttons, button holes, decorations (embroidery, etc).
- No peeling off of coating from beads or decorations. Do not report the numerical rating for non-textile materials.

- No noticeable puckering at seams and other portions
- Label Assessment
- Dimensional change is not to be assessed in appearance test.

Colour fastness to bleach

- Colour change ≥ 4

Colour fastness to drycleaning

- Colour change ≥ 4
- Self staining ≥ 4.5

Ironing

- The appearance of sample after ironing should be satisfactory.
- If ironing is applied for improving the appearance then the same should be mentioned in the reportas follows:

'Cool iron if necessary' or 'warm iron if necessary' or 'warm steam iron if necessary' or 'hot iron if necessary'

- If the sample does not meet the requirement of shrinkage after wash but meets the same after ironing then mention as follows:

'Cool iron' or 'Warm iron' or 'hot iron' or 'Warm steam iron'.

4.3.7.6 Procedure

Washing and drying
Perform the colour fastness to washing test at m/c wash warm and at cold. Wash sample at appropriate wash and dry conditions depending on fibre composition and product type (Table 4.2).
Assess the results of colour fastness to wash and dimensional stability and appearance.
If sample passes for m/c wash warm then recommend this. If sample fails for m/c wash warm and tumble dry conditions then proceed for either m/c wash warm, line dry or less severe conditions, i.e., m/c wash cold and line dry/flat dry. If sample fails for m/c wash then proceed for hand wash.
Perform colour fastness to drycleaning test only in normal styles. In embellished/delicate styles, perform appearance after drycleaning test also.
Drying method should be selected keeping in mind the style of merchandise.

Bleaching

US market

- If sample passes for both chlorine and non-chlorine bleach then no need to mention any instruction for bleach.

- If sample passes for non-chlorine bleach only then mention
- 'Only non-chlorine bleach when needed'
- If sample fails for both bleaches then mention
  'Do not bleach' or 'No bleach'

UK and European market

- Perform colourfastness to chlorine bleach for whites only. If sample passes then recommend'Only chlorine bleach when needed' otherwise mention 'Do not bleach'.
- For coloured goods, always mention 'Do not bleach'.

Ironing

Select the ironing temperature depending upon fabric type. Refer guideline given in the Table 4.2. Assess the appearance of sample after ironing.Drycleaning If sample passes for colour fastness to drycleaning using perchloroethylene then recommend

'Dry clean' and Symbol 

If sample fails for drycleaning then report 

Additional care instructions:

- Wash inside out: Where garment having heavy/delicate embellishments.
- Iron on reverse: It is to be recommended for garments having ornamentation such as beads and sequins or prints (e.g. plastisol, flock, etc.) or other non-textile material where ironing on such parts can deteriorate the appearance.
- Dry away from direct sunlight/dry in shade: It should be generally recommended in case of silk fabrics and pastel colour fabrics.
- Steam iron recommended: Where puckered appearance occurs mostly in double stitching or around cut work/applique.
- Steam iron also improves appearance in case of cellulosic materials by removal of crinkles.
- Iron whilst damp: Where dimensional stability improves on iron whilst damp (s.a. moss crepes, georgettes, etc.)
- Reshape whilst damp: Can be recommended for moss crepes, chiffon, knits depending on product style.

Note:

- Recommend washcare after three washes for USA and after one wash for the UK and Europe.
- Dry clean after one wash
- Always mention dry clean as one word (Dry Clean).

4.3.7.7 Reporting instruction

Results of all test parameters and requirement should be given on the report.

If sample fails for both washing and drycleaning then conclusion to be given as 'Care label cannot be recommended unless the failed parameters are improved'.

If any of the instruction with respect to temperature in washing, ironing and drying is not given, it indicates that the product is suitable for severest condition.

Conclusion is to be given as:
Based on the results of the tests reported, proposed care instructions for the submitted sample may read as:

'Symbols are to be given'

'Instruction in wordings to be given'

In case of the UK/Europe buyers where manual specify wash care on composition basis. If sample fails that specified instructions,we can recommend the instructions with a comment :

'Recommended wash care deviates from the manual specified instructions'.

=> In case the care label is recommended at fabric stage, the following comment should also appear below the recommended care instructions.

However, further tests on a garment made of submitted sample for confirmation is recommended.

Special Cases:

1. Crinkled styles with twist, wring or crush effect:
   Dimensional stability is to be checked on plain fabric. Drying instructions are to be taken from client.
2. Damp wipe only:
   Can be recommended for non-washable and non-dry cleanable products such as mats with PVC coating, wooden mats, mats with special effect, etc.
3. For towels/bath mats dry clean only should not be recommended if it fails for hand wash and m/c wash.
4. It is advisable to not to recommend 'Dry clean only' care instructions for cotton fabrics and inner wears if the sample fails all the washing criteria. The following comment can be given:
   'Washing instructions cannot be recommended however the submitted sample meets the Drycleaning instructions'. All the results machine wash, hand wash and dry clean should be reported.

4.4 Fibre products identification

4.4.1 The textile fibre products identification act

The Act was passed 'to protect the public against misbranding and false advertising' of textile articles. This act (22) is under the jurisdiction of the Federal Trade Commission. As per the Act, labels must be securely affixed to the garment where they will be conspicuous to the consumer at the time of sale. Required information in the label includes:

i. Generic name and percentage of all fibres in amounts of 5% or more
ii. Name of manufacturer or registered identification number
iii. Country o.f origin

Fibre content may appear on the reverse side of the label if label states: 'Fibre content on reverse side'. Here, it is to be noted that failure in fibre content affects quota category, duty rate and labelling compliance.

Under this Act, the generic name, and percentage of all fibres in amounts of 5% or more, must be listed in predominance by weight. Fibre trade names may be used in conjunction with generic name but may not be used exclusively. Fibres present in amounts less than 5% cannot be identified by their generic names but should be labelled as 'Other fibres' unless the fibres have functional significance, such as 96% cotton and 4% spandex for elasticity.[22]

No tolerance exists for products made wholly of one fibre. In such a case, product should be labelled as '100%' or 'All'. However, there is a ±3% tolerance, by weight, for products composed of more than one fibre. For instance, a product of 55% cotton and 45% polyester can be deviated as 58% cotton and 42% polyester or 52% cotton and 48% polyester.

In addition, there are other aspects which are also to be considered when there is ornamentation in the garment. 'Exclusive of Ornamentation' may be used when ornamentation (fibre and yarn) incorporated into the fabric for aesthetic appeal does not exceed 5% of the total fibre weight. There is another terminology, i.e. 'Exclusive of Decoration' which is to be used when trimmings such as embroidery and appliqués do not exceed 15% of the garment surface area.

As per the updated children safety guideline, monofilament thread is not allowed to use on apparel for infants and children under 6 years of age for few brand buyers of USA. This is mainly due to protection from stretch and possible strangulation, which may be happened in such garments while in use by such class of children.

4.4.2 Wool products labelling act

Under the general requirements as mandated by the United States Federal Trade Commission in16 CFR 300, the Wool Products Identification Act was established 'to protect producers, manufacturers, distributors and consumers from the unrevealed presence of substitutes and mixtures in spun, woven, knit, felted, or otherwise manufactured wool products....'

The required information, which should be available to the consumer at the time of sale, includes amount of wool in exact percentage by weight, manufacturer's registered identification number (RN) or wool labelling number (WPL) and country of origin.

Generic name percentage of all wool fibres should be listed in predominance by weight. Trade names of fibres may be used in conjunction with generic name but may not be used exclusively. However, the following speciality fibres may be used instead of the word 'wool': Mohair, Cashmere, Camel Hair, Vicuna, Llama, Alpaca fibres (other than wool) present in amounts less than 5% cannot be identified by their generic names but should be labelled as 'other fibres' unless the fibres have functional significance. However, no tolerances exist for products made wholly of wool. In this case, the product should be labelled as '100%' or 'All'.

The term 'virgin' or 'new' can only be used to describe a fibre that has never been reclaimed from a product previously spun, knitted, woven or otherwise made into a textile product. Recycled wool, which has been woven or felted into a wool product and then returned to a fibrous state without having been used by the ultimate consumer, must be identified.

4.4.3 The textile products (indications of fibre content) regulations

This is the requirement for export destined to United Kingdom and European community (5). Under this new regulation,[23] fibres must be listed by generic name and predominance by weight.

Products made entirely of one fibre should be labelled as '100%', 'Pure', or 'All'. This is also allowed when:

There are other fibres less than 2% of the total weight resulting from inadvertent impurities during manufacture,

There are other fibres less than 7% of the total weight which are visible and distinct and intended to produce a decorative effect and

There are other fibres less than 2% of the total weight incorporated for an antistatic effect.

However, there is 3% tolerance, by weight, for products composed of more than one fibre.

In case of products where one fibre is at least 85% of the total weight, the following is the guideline:

Fibre name and percentage. Example: 90% cotton; or,

Fibre name and percentage followed by a minimum. Example: 85% cotton minimum; or,

Fibre name and percentage in predominance by weight. Example: 85% cotton/15% polyester.

The guideline differs in the products where no fibre is at least 85% of the total weight and is given below:

Fibre name and percentage in predominant order by weight; or,

Fibre name without the percentage in predominant order by weight

Fibres making up less than 10% of the total fibre weight may be listed as 'Other fibres' as long as percentage weight is indicated; or,

Fibres making up less than 10% of the total fibre weight may be listed separately by fibre name and percentage.

4.4.4   Products made of 'fleece wool' or 'virgin wool'

These terms may only be used in the product made entirely of wool fibre that has not been made into any previously finished product, or has been through any spinning or felting operations, or has been damaged in any other manufacturing process. This is also allowed when the product contains multiple fibres, and fleece or virgin wool makes up at least 25% of the total weight, and when there is scribbled mixture, the mixture contains only fleece or virgin wool and one other fibre.

References

1.   Consumer products safety commission, General information. Available from: http://compliance.alternativeapparel.com [Accessed 2 March 2009].
2.   Title 16 CFR part 1610 Standard for the flammability of clothing textiles.
3.   North central regional extension publication (2003), 'Facts about fabric flammability', 174, 1–7.
4.   16 CFR parts 1615/1616 Standard for the flammability of children's sleepwear: sizes 0 through 6X; Standard for the flammability of children's sleepwear: sizes 7 through 14.
5.   Bureau Veritas Consumer Products Services (2002), 'Open The Door to Quality', 21–46.
6.   The Nightwear (Safety) Regulations 1985.
7.   British Standard 5722 Flammability performance of fabrics and fabric assemblies used in sleepwear and dressing gowns.
8.   British Standard 5438 Methods of test for flammability of vertically oriented textile fabrics and fabric assemblies subjected to a small igniting flame.
9.   British Standard 5651 Cleaning and wetting procedures for use in the assessment of the effect of cleansing and wetting on the flammability of textile fabrics and fabric assemblies.
10. The Nightwear (Safety) (Amendment) Regulations 1987, ISBN 0110785134.

11. BS EN 14878 Textiles – Burning behaviour of children's nightwear.
12. BS EN 1103Textiles. Burning behaviour. Fabrics for apparel. Detailed procedure to determine the burning behaviour of fabrics for apparel.
13. Das Subrata (2005), 'Care labels: Some truths from Bangla apparels export', *Indian Text J*, 115, 84–87.
14. 16 CFR part 423 Trade regulation rule on care labeling of textile wearing apparel and certain piece goods.
15. ISO/FDIS 3758:2003(E) Textiles -- Care labelling code using symbols.
16. Das Subrata (2005), 'Studies on care issues of high performance apparels', *Express Textile*, 9, 29.
17. ASTM D 3136 Terminology relating to labels for textile and leather products other than textile floor coverings and upholstery.
18. ASTM D5489-01a Standard guide for care symbols for care instructions on textile products.
19. Das Subrata (2005), 'A study on the performance and prospect of readymade garments in Bangladesh', *Pakistan Text J*, 54, 53.
20. Das Subrata (2006), 'Wash care fundamentals: a never ending discussion, part I', *Apparel Views*, 5, 52–54.
21. Das Subrata (2006), 'Wash care fundamentals: a never ending discussion, part II', *Apparel Views*, 5, 58–60.
22. 16 CFR 303 Rules and regulations under the textile fiber products identification act.
23. The Textile products (indications of fibre content) (amendment) regulations 2008, No.6.

CHAPTER 5

Safety issues for different accessories in children garment

Abstract

With a view to avoid potential hazards and to comply with safety standards, components used in children apparel such as zippers, drawstrings, fasteners, and decorative attachments, etc., must not pose any harm to children during normal use, or as a result of any foreseeable damage or abuse. The chapter discusses the importance of safety issues of different accessories used in children's apparel. The chapter also discusses possible hazards associated with the application of different accessories in children garment and the guideline on the standards of such items for safe use.

Keywords: hazards, children's apparel, fasteners, drawstrings, bows

Chapter contents (Section headings)

5.1   Importance of safety issues

Brand buyers are committed to provide quality products. The need for safety is crucial and is strongly recognized across all markets, especially for children's cloths.[1] The right guideline shall assist the suppliers with safety through the product development process. It will help to establish the required safety standards and produce apparel that will limit potential hazards. The information, legal legislation and regulations are intended to assist with manufacturer's legal obligations and produce a product that meet and in some cases, exceeds all the legal and industry requirements across all markets worldwide. Strict adherence to these standards is not optional. If a supplied product does not meet all the required safety standards, laws, rules, or regulations, a supplier would be liable to the brand for product withdrawals/ recall costs and customer returns to store along with other legal obligations under the agreement and applicable laws. Components used on children's apparel, such as zippers, drawstrings, fasteners, and decorative attachments must not contain any hazards or any hazards as a result of any foreseeable damage or abuse   during normal use.

In view of the above facts, different accessories used in children's apparel have been discussed along with the possible hazards associated with their potential application and the guideline on the standards of such items.

5.2   Small parts: choking hazards

Children aged 3 years and under are particularly susceptible to choking, asphyxiation and ingestion hazards caused by small objects.[2] All components that could be detached from children's clothing are all examples of small parts, and, thus, cause choking. Some examples of small parts are:snaps/studs/rivets; buttons; appliqués; bows and rosettes; pom-poms and fringe; dungaree clasp (Hasps) and slider; zipper components; belt fastenings; toggles; decorative and functional loops; and decorative labels.

If the trim or component can fit within the small parts cylinder (Fig. 5.1), the item is considered as a potential choking hazard. It is, therefore, a general policy in the US, where buyers should ensure that for all small parts intended for children aged 3 years and under should withstand a 15 to 21-lb pull force.

5.3   Metal fasteners

All metal fasteners (including any surface coatings) [Fig. 5.2] such as press fastener (prong), press fastener (post), stud button, eyelet, and rivet must be

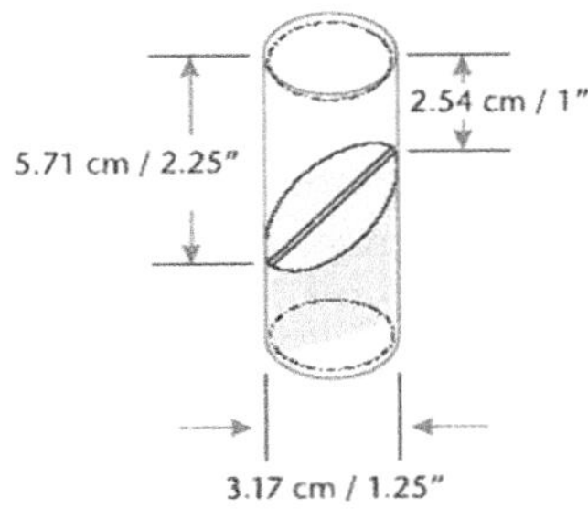

Figure 5.1 Small parts cylinder

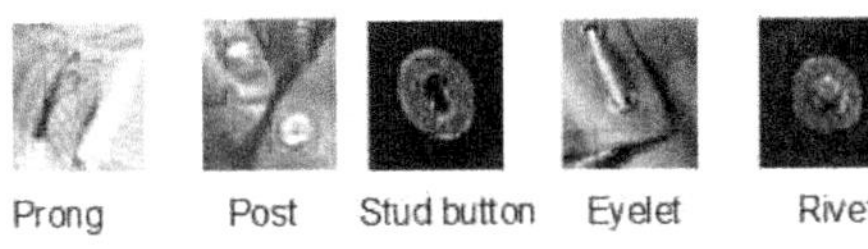

Figure 5.2 Metal fasteners

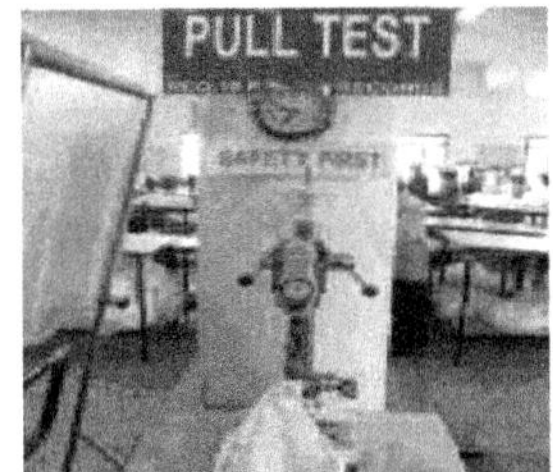

Figure 5.3 Pull test equipment

only from an approved source like YKK, Prym, Scovill Fasteners Inc., or Morito (Kane-M). In order to ensure metal fasteners are securely attached to garments, the minimum pull force requirements of 15–21 lbs measured on pull test equipment (Fig. 5.3) must be achieved depending on the requirement of a buyer.

Metal fastenings on all children's product must not contain toxic elements. For the US, this must include those toxic elements specified as in ASTM F963[3] and for Canada, those toxic elements specified as in The Hazardous Products Act.[4] If the component has surface coating, it must comply with the lead requirements outlined in CFR, Title 16, Part 1303[5] for USA and Hazardous Product Act as amended on April 19, 2005 for Canada. All metal fastenings must comply with The European Nickel Directive (94/27/EC).[6] Recently announced CPSIA 2008[7] is now applicable for such items with regard to quality assurance.

The fastener and coating must be capable of withstanding washing and drycleaning in accordance with the garment care label. All press fasteners including stud, post, socket, and cap must be metal and must be non-ferrous to ensure garments can pass through the metal detector. This includes metallic finishes. Fasteners must be free from rust, contamination, oxidation, and all other types of degraded corrosion.

5.4   Zipper fasteners

Zipper must be sourced from an approved supplier as a complete unit with all of the necessary components such as top stop, slider, bottom stop, etc. Zipper components must not be individually purchased and self assembled.

For the UK, zipper and zipper pulls must comply with BS 3084:2006.[8]

For USA and Canada, zippers must comply with ASTM D 2060[9] and D 2061,[10] respectively.

For Germany, zippers must comply with DIN 3419-1[11] and for Japan, zippers must comply with JIS standard.[12]

In order to ensure zippers and zipper pulls are securely attached to garments, the pull force requirement of 15 lbs is required.

For USA and Canada, zippers used on clothing for children aged under 3 years must conform to the torque and as per the specifications in ASTM F963 and CFR Title 16-Part 1500,[13] respectively. Some of the specifications are as follows:

- 0–18 months   2 ± 0.2 lbf. in. (0.23 Nm)
- 18–36 months   3 ± 0.2 lbf. in. (0.34 Nm)
- 36–96 months   4 ± 0.2 lbf. in. (0.45 Nm)

Zippers on all children's product must not contain toxic elements. If the zipper pull has a surface coating, it must comply with the CPSIA 2008. All zippers and zipper pulls must conform to the European Nickel Directive and must be non-ferrous to ensure garments pass through the metal detector.

Zippers must have full autolock or semi-autolock sliders. Pin lock zippers are not acceptable on children's wear.

Top edges of the zippers are to be finished properly such that there are no sharp edges on the teeth or top stops. Fasteners cannot have rough or sharp edges and they must be free from rust, contamination, oxidation, and all other types of degraded corrosion.

Zipper stops are specially designed so that the zipper slider cannot be removed in children's clothing. Coil zippers for children's clothing must have moulded plastic top and bottom stops. For USA and Canada, metal zippers

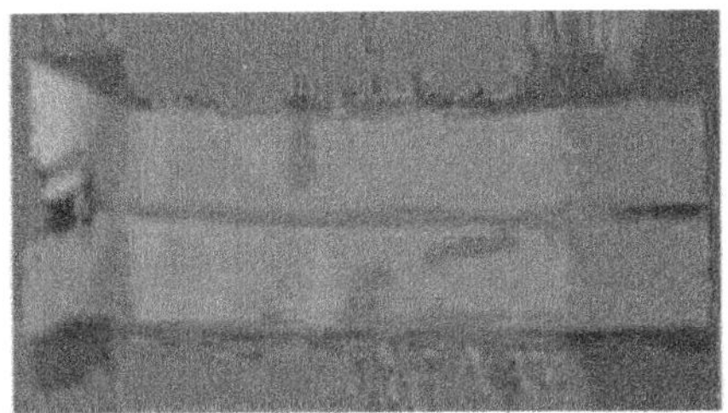

Figure 5.4   Invisible / concealed zipper

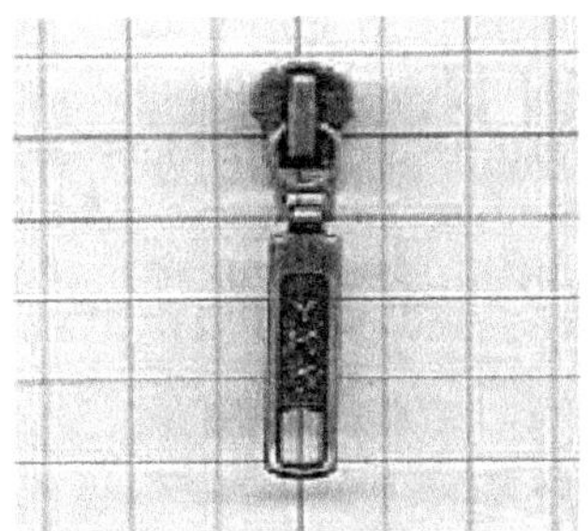

Figure 5.5   Two-piece zipper pulls

require a zipper guard or facing to prevent the zipper being in direct contact with the skin. However, invisible/concealed zippers (Fig. 5.4) are not permitted in the clothing for children aged 3 years and under, i.e., sizes 0–5T for USA and sizes 0–3x for Canada.

A zipper pull is the component used to open and close a zipper. It can be attached through either the main body of the zipper slider or through the eye of the zip pull. Indirect attachment pulls (two-piece zipper pulls) [Fig. 5.5] are not allowed in the clothing for children aged 3 years and under.

Ring pulls or other open-type designs (Fig. 5.6) are not permitted for in the clothing for children aged 3 years and under.

5.5   Dungaree clips (hasps) and sliders

All dungaree clips (hasps) and sliders (including any surface coatings) [Fig. 5.7] must be only from an approved source like YKK, Prym, Scovill Fasteners Inc., or Morito (Kane-M). In order to ensure dungaree clips (hasps) and sliders are securely attached to garments, the minimum pull force requirements of

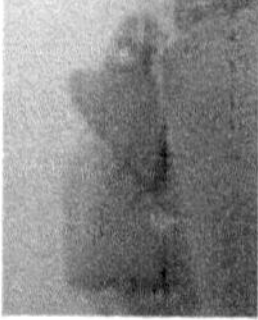

Figure 5.6   Ring pulls or other open type designs

15–21 lbs measured on pull test equipment must be achieved depending on the requirement of a buyer.

Dungaree clips (hasps) and sliders on all children's product must not contain toxic elements. For the US, this must include those toxic elements specified in ASTM F963 and for Canada those toxic elements specified in The Hazardous Products Act. If the component has surface coating, it must comply with the lead requirements outlined in CFR, Title 16, Part 1303 for USA and Hazardous Product Act as amended on April 19, 2005 for Canada.

Figure 5.7   Dungaree clips (hasps) and sliders

Dungaree clips (hasps) and sliders must comply with The European Nickel Directive (94/27/EC). Recently announced CPSIA 2008 is now applicable for such items with regard to quality assurance.

Dungaree clips (hasps) and sliders and coating must be capable of withstanding washing and drycleaning in accordance with the garment care label. They must be metal and must be non-ferrous to ensure garments can pass through the metal detector. This includes metallic finishes also. They must be free from rust, contamination, oxidation and all other types of degraded corrosion.

5.6   D-rings

D-rings (including any surface coatings) [Fig. 5.8] must be only from an approved source like YKK, Prym, Scovill Fasteners Inc., or Morito (Kane-M). In order to ensure D-rings are securely attached to garments, the minimum pull force requirements of 15–21 lbs measured on pull test equipment must be achieved depending on the requirement of a buyer.

D-rings on all children's product must not contain toxic elements. For the US, this must include those toxic elements specified in ASTM F963 and for Canada those toxic elements specified in The Hazardous Products Act. If the component has surface coating, it must comply with the lead requirements outlined in CFR, Title 16, Part 1303 for the US and Hazardous Product Act as amended on April 19, 2005 for Canada. D-rings must comply with The European Nickel Directive (94/27/EC). Recently announced CPSIA 2008 is now applicable for such items with regard to quality assurance.

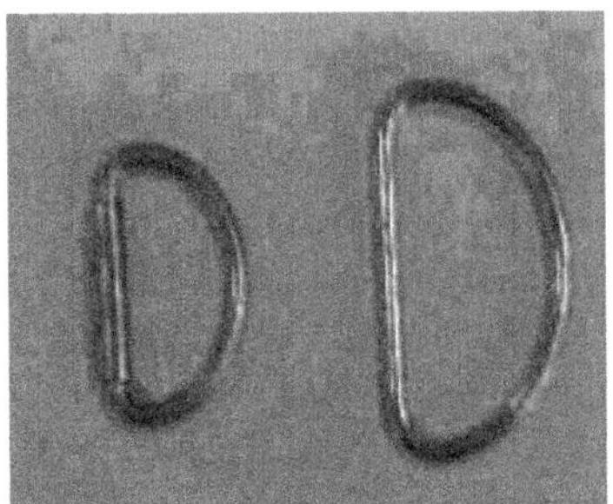

Figure 5.8   D-rings

D-rings and coating must be capable of withstanding washing and drycleaning in accordance with the garment care label. They must be metal and must be non-ferrous to ensure garments can pass through the metal detector. This includes metallic finishes also. They must be free from rust, contamination, oxidation, and all other types of degraded corrosion. D-rings are not allowed on sizes 0–24 months and are not permitted at the free ends of ties.

Must have no visible joins, designed so that they cannot be detached from the garment, and secured close to the garment.

–Sizes 2T – 5T	maximum inside diameter: ½″
–Sizes	
Girls: 4-6X	maximum inside diameter: 1″
Boys: 4-7	
–Sizes	
Girls: 7-16	
Boys: 8-18	maximum inside diameter: 1½″

5.7 Functional and non-functional drawstrings, cords/ ties

In order to ensure functional and non-functional (decorative) cords/ties and drawstrings are securely attached to children garments, a minimum pull force of 15 lbs is required. Cord / ties made from metal chain must not be used for children aged 3 years and under. For children over 3 years, cords/ties/belts made from metal chain must only be non-functional (decorative) and have a breakaway strength of 25 N/5.6 lb to 40 N/9 lb. The length of the cord/tie is dependent on area of attachment. Children's garments must not be designed to have functional drawstrings, non-functional (decorative) drawstrings, functional cords/ties or non-functional (decorative) cords/ties which emerge from the back of the garment and they must not be tied at the back of the garment. Sashes, however, are allowed to be tied at the back of the garment. The ends on all drawstrings and cords/ties must be secured with either, a double turn secured with lockstitch, a heat seal, laser cut, or a plastic sleeve (shoe lace end). If a plastic sleeve is used, it must withstand a 100-N/22.5-lbs pull test. The free ends of all drawstrings and cords/ties must not be secured with a knot or equivalent, i.e., bead, toggle, or pom-pom for children's clothing. Beads must not be used on functional or non-functional (decorative) drawstring and cords/ties for children aged 3 years and under. For children above 3 years, beads can be used on drawstring and cords/ties, except when positioned at the free ends. Functional drawstrings, non-functional (decorative) drawstrings, functional cords/ties and non-functional decorative cords/ties (Fig. 5.9) used on the lower edges of a garment must not hang below the hem of the garment. There are also maximum length restrictions to consider when using drawstrings and cords/ties on clothing for children.

No functional drawstrings, cords, or ties are allowed in any children's size range in the hood or neck area of garments. Elastic cords/ties are not permitted in the hood and neck area of garments. However, non-functional drawstrings and non-functional (decorative) cords/ties are only allowed on the hood at

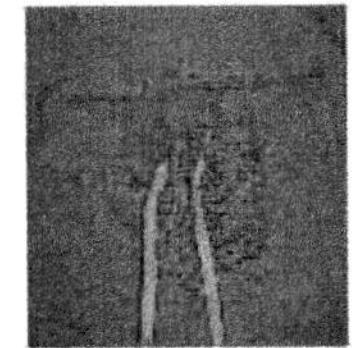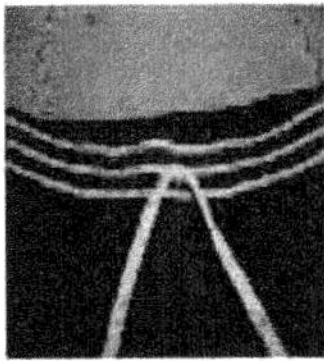

Figure 5.9   Drawstrings & cord / tie

the base of the front opening. Non-functional drawstrings are to be securely attached with a bartack 1cm/½″ from the exit point.

In case of waist area, functional drawstrings must be secured to the garment with a bartack to prevent the drawstrings from being pulled out of the garment. This also prevents one end of the drawstrings from ending up longer than the other and thus becoming an entrapment/catch hazard. The most common place to secure the drawstring is at the centre back waist.

Non-functional drawstrings should exit through button holes or eyelets at the waist. The cord must be secured with a bartack no further than 1 cm/½″ from the buttonhole or eyelet. Non-functional (decorative) cord/ties should be attached to the outside of the garment with a bartack. Measurement is taken from the point of attachment. Lace up ties must be securely attached at exit points for 3 years and under, and at midpoint for all others.

The US Consumer Product Safety Commission (CPSC) approved a new federal safety rule for drawstrings in children's outerwear on 1st July 2011. The rule designates children's upper outerwear in sizes 2T through 12, with neck or hood drawstrings, and children's upper outerwear in sizes 2T through 16, with certain waist or bottom drawstrings, as substantial product hazards.

As per the statistics of European countries,[14] serious accidents involving cords and drawstrings on children's clothing fall into two main groups by age of child:

i. Younger children: Persons aged from birth to age 7 years (that is 6 years and 11 months) which includes all children up to and including a height of 134 cm — Entrapment of hood cords in playground equipment such as slides, resulting in fatalities.

ii. Older children and young persons: Persons aged from 7 years up to age 14 years (that is up to 13 years and 11 months) which includes all boys of height greater than 134 cm up to 182 cm and girls of height greater than 134 cm up to 176 cm — entrapment of cords and strings from the waist and lower hems of garments in moving vehicles such as "bus doors", ski lifts, and bicycles resulting in severe injuries or death from being dragged along or run over by the vehicle.

iii. Hook and neck area on garments for young children: Garments intended for young children shall not be designed, manufactured or supplied with drawstrings, functional cords or decorative cords in the hood or neck area.

Hook and neck area on garments for older children and young persons: When the opening of the garment is at its largest and the garment is laid flat there shall be no protruding loop. When the garment opening is at its smallest, that is the size it is intended to fit, the maximum protruding loop circumference shall be 150 mm. Functional cords shall not be more than 75 mm in length at either end and shall not be made from elastic cords. Decorative cords shall not be more than 75 mm in length at either end including any attachment such as toggle and shall not be made from elastic cords. Halter neck style garments shall be constructed with no loose ends in the hood and neck area.

iv. Waist area of garments: Drawstrings in the waist area shall protrude by a maximum of 140 mm at each end when the garment is flat on pattern and by no more than 280 mm when closed to the intended waist size. Functional cords and decorative cords in the waist area shall be a maximum of 140 mm including any embellishment on decorative cords. Belt loops shall be designed to lie flat against the garment. Sashes shall be acceptable provided that when untied they do not hang below the hem of the garment. The length of sash when untied measured from the point where it is to be tied shall be no more than 360 mm.

v. Lower hems of garments which hang below the waist: Drawstrings, decorative cords or functional cords including any toggle on the lower edges of garments where the lower edge is situated below the hip shall not hang below the lower edge of the garment and should be totally inside the garment. If they are external to the garment, the drawstring or cord shall lie flat against the garment, when the garment is tightened or fastened. There shall be no protruding drawstrings, functional cords or decorative cords on the bottom hem of coats, trousers or skirts, which are designed to finish at the ankle.

vi. Sleeves: Drawstrings, functional cords and decorative cords at the lower edge of long-sleeved garments shall be totally on the inside of the garment, when the garment is fastened. However, they are acceptable on short-sleeved garments provided the sleeve finishes above the elbow and the maximum protruding length is 140 mm measured laid flat on pattern.

vii. Other parts of the garment: In all the other areas of the garment, not previously addressed, the drawstrings or functional and decorative cords shall protrude by no more than 140 mm when the garment is open to its largest.

Based on the case studies, two types of hazards are recognized in regard to the use of drawstrings in children's garments:

a) the potential strangulation hazard primarily associated with hood and neck drawstrings, and

b) the potential vehicular dragging hazard primarily associated with waist and bottom drawstrings.

5.7 Toggles

In order to ensure toggles (Fig. 5.10) are securely attached to children's garment, a minimum pull force of 15 lbs is required. Wood, cork, leather, mother of pearl (shell), glass, or other non-durable toggles must not be used on children's clothing. Toggles must only be used on functional drawstrings, non-functional (decorative) drawstrings, functional cords/ties or non-functional (decorative) cords/ties that have no free ends. In addition, free ends cannot be knotted together to form a continuous loop. Positioning of toggles on children's clothing is to be seriously viewed in order to prevent injury and discomfort during wear. For instance, toggles positioned at the knee area are not acceptable.

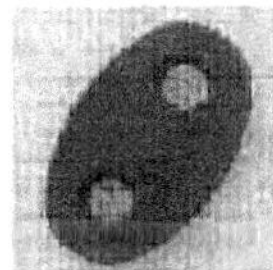

Figure 5.10 Different types of toggles

i. Decorative purpose: Pig nose (cord lock) and spring loaded toggles can be used on children's clothing as a decorative application. The cord and toggle must sit flush to the garment and must be securely contained within the loop of the cord so it cannot be removed from the garment. The cord must be securely bartacked no more than 1 cm/3/8″ from the eyelet or button hole.

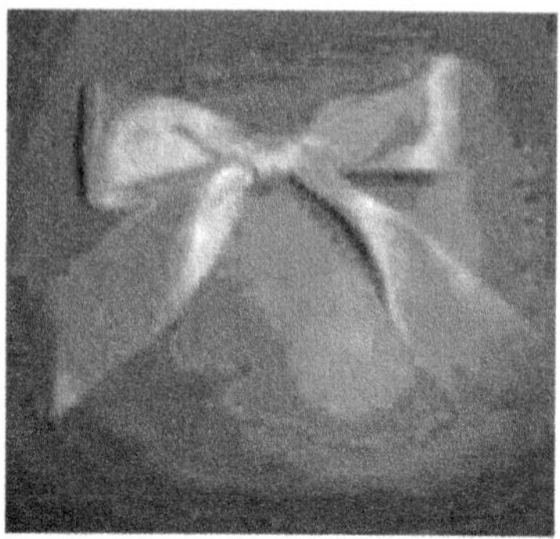

Figure 5.11   Fixed bows

ii. Functional purpose: Pig nose (cord lock) and spring loaded toggles can sometimes be used on children's clothing as a functional application. The loop circumference cannot extend more than 7.5 cm (3″) when cord is cinched to the body and must fully retract inside waistband when the garment is fully extended. The toggle cannot be attached to a continuous drawstring.

5.8   Fixed bows

In order to ensure fixed bows (Fig. 5.11) are securely attached to garments a minimum pull force requirement of 15 lbs must be achieved for children aged 3 years and under. Fixed bows must be tested for colour fastness to water. Bows must be secured at the centre with a bartack. The tail ends of the bow must be secured with either a double turn secured with a lockstitch, a heat seal or laser cut. Fixed bows can be used on all areas of a garment. However, there are maximum length restrictions to consider for the loop and tail lengths of bows when used on children's clothing. For instance, in the hood area, loop (4 cm) and tail (2.5 cm) differs from waist area wherein accepted values are 7.5 cm and 7.5 cm, respectively.

5.9   Buttons

Wood/cork/leather/mother of pearl (shell)/glass or other non-durable buttons must not be used on children's clothing. Buttons cannot have rough or sharp edges and must be free from rust and contamination. They should not contain toxic elements, objectionable surface coating, and must comply with the CPSIA 2008. With a view to ensure buttons are securely attached to the garment, a minimum pull force of 15 lbs is required for children aged 3 years and under. Two-piece multi-component buttons and fabric-covered buttons must not be used on children aged 3 years and under.

All buttons must be attached using a lockstitch attach machine. Two-hole buttons must have 14–16 stitches and four-hole buttons must have 24–26 stitches. Four-hole buttons must be stitched through each hole and only core spun polyester sewing thread to be used while stitching buttons. Bobbin thread should be a different colour than the needle thread, which enables easy verification of the lockstitch application. For buttons that require a harsh wash, it is recommended that buttons are attached after washing to avoid damage to either the button or garment.

5.10   Pom-poms and fringe

Traditional pom-poms and fringe (Fig. 5.12) made from hand knitting/sweater yarns and those constructed with metal components are not permitted for children aged 3 years and under. Stuffed pom-poms made from fabric are acceptable. The filling must be new and completely enclosed inside a lining. In order to ensure secure attachment to garments, a minimum pull force of 15 lbs is required. Pom-poms and/or fringe must not be attached to the end of a drawstring, cord or tie with free ends.

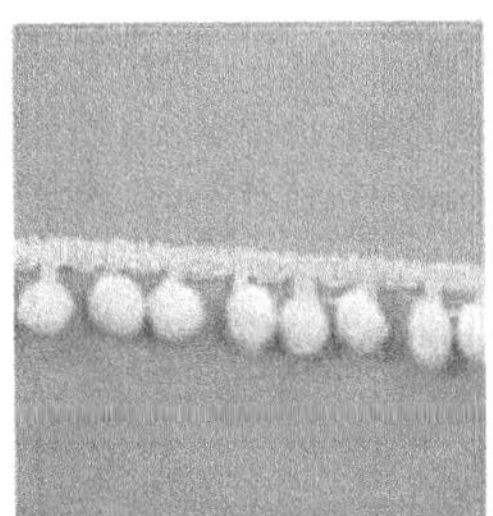 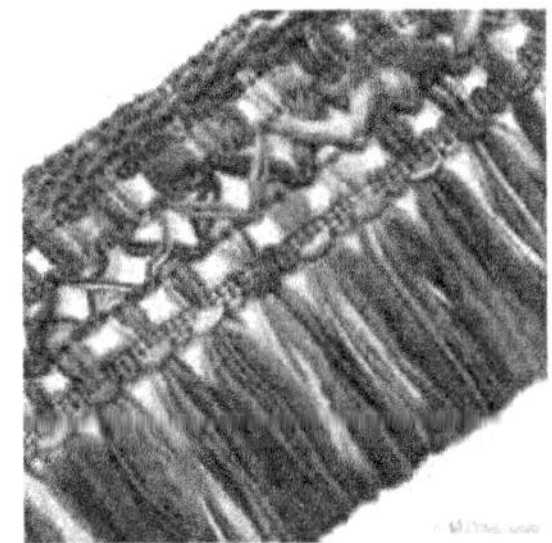

Figure 5.12   Pom-poms and fringe

5.11   Decorative trims and embellishments

Appliqué and Embroidery: Appliqué can be edge stitched or stitched at the centre (Figs. 5.13a, b). In case of centre stitched, it must be lockstitch attached and should withstand a minimum pull force of 15 lbs.

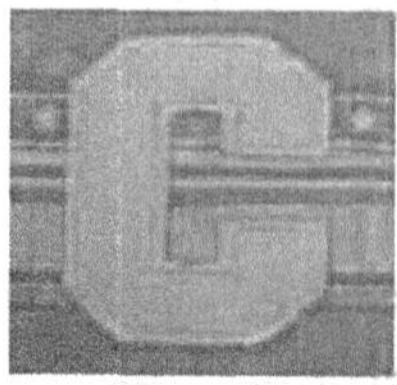 

Figure 5.13   Applique: a) edge stitched; b) center stitched

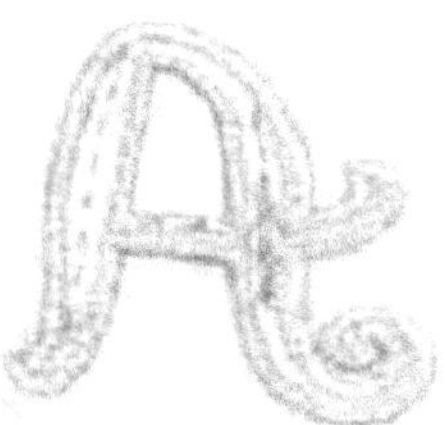 

Figure 5.14   Examples of embroidery

Embroidery (Fig. 5.14) is to be backed with interlining if the reverse is scratchy and comes in direct contact with the skin. If backing is utilized it must be permanently attached so that it cannot pose a potential choking hazard.

5.12 Bead

Individual beads (Fig. 5.15) can be stitched by hand for 4 years and above by using core spun polyester. Those must be securely attached with double thread and the end of thread is to be knotted. Maximum thread end of 1 cm (3/8″) and minimum 0.5 cm (3/16″) are acceptable but floats over 1 cm (3/8″) are not acceptable. Beads should not contain toxicity or any undesirable surface coating. There should not be any loss in colour, loss of bead and peeling or delamination after wash or drycleaning.

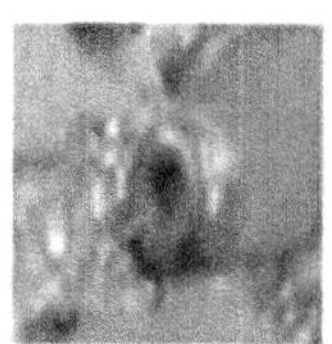  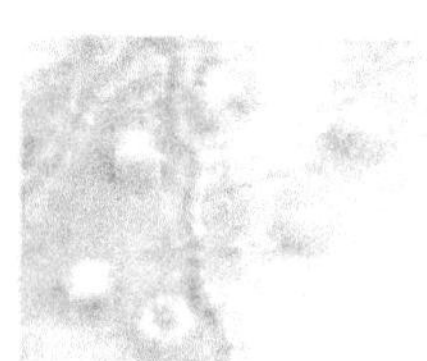

Figure 5.15   Examples of bead

5.13   Sequins

Individual sequins (Figure 5.16) can be attached in one part of garment (1 year and above) or all over the garment (4 years and above) by hand. Those must be securely attached with double thread and the end of thread is to be knotted. Sequins can also be attached by machine, individually (1 year and above) or in a row (4 years and above).

They must be securely lockstitch the sequins and secured. In both hand and machine attachment, maximum thread end of 1 cm (3/8″) and minimum 0.5 cm (3/16″) are acceptable, but floats over 1 cm (3/8″) are not acceptable. Sequins should not contain toxicity or any undesirable surface coating. There should not be any loss in colour, loss of sequin and peeling, or delamination after wash or drycleaning.

5.14   Jewel

A jewel (Fig. 5.17) is considered to be greater than 5 mm. It is considered a bead if less than 5 mm. Individual jewels can be attached to the garment by hand for children's at an age of 4 years and above. Jewels must have attachment holes at each side. It is attached by stitching at least three times at either side.

Those must be attached with double thread and the end of thread is to be knotted. Maximum thread end of 1 cm (3/8″) and minimum 0.5 cm (3/16″) are acceptable, but floats over 1 cm (3/8″) are not acceptable. Jewel should withstand a minimum pull force of 15 lbs. Jewels should not contain toxicity

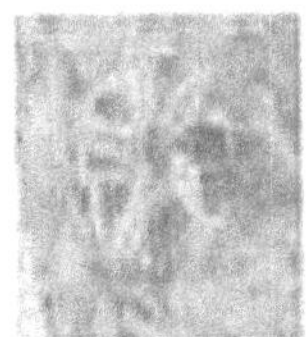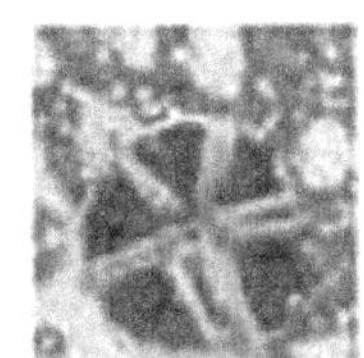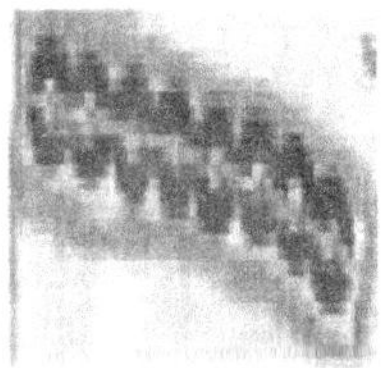

Figure 5.16   Examples of sequins

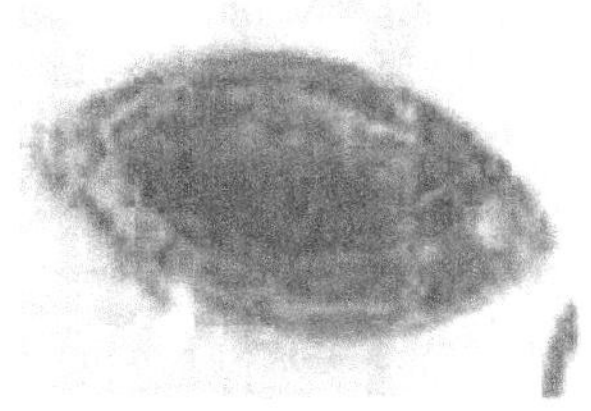

Figure 5.17   Example of jewel

or any undesirable surface coating. There should not be any loss in colour, loss of sequin and peeling, or delamination after wash or drycleaning.

5.15   Heat transfer diamante (rhinestone) and stud guidelines

The rhinestone and stud are suitable for children of age 4 years and above (Figure 5.18). Right temperature, time and pressure are to be maintained for secure attachment.

They should not contain toxicity or any undesirable surface coating. There should not be any loss in colour, loss of diamante (rhinestone), or metal stud and peeling or delamination after wash or drycleaning.

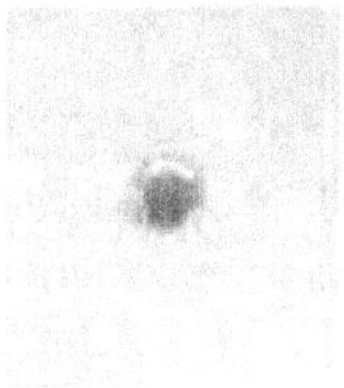 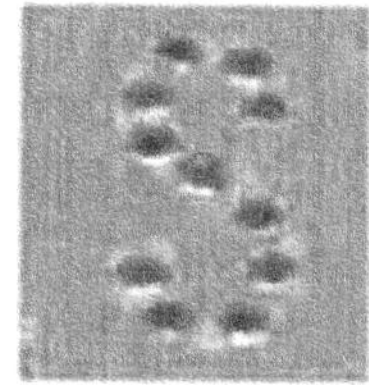 

Figure 5.18   Examples of heat transfer diamante (rhinstone) and stud

References

1.   TH – CTS (2001), Tommy Hilfiger manual, Safety statement, 2.
2.   Safety manual (2005), Wal-Mart / George technical manual, 7.
3.   ASTM F963 Standard consumer safety specification for toy safety.
4.   Hazardous Products Act (2008), Canadian center for occupational health and safety. Available from: http://www.oshforeveryone.org/leg/documents/canada/caehpa/caahaze0.htm [Accessed 11 March 2009].
5.   16 CFR part 1303 Ban of lead containing paint and certain consumer products bearing lead containing paint.
6.   The European parliament and of the council directive 94/27/EC(1997), 'The European directive restricting the use of Nickel'. Available from: http://www.teg.co.uk/teg/nickel/94-27-EC.htm [Accessed 11 March 2009].
7.   CPSIA 2008: Consumer product safety improvement act of 2008, Public law 110-314. Available from: http://www.cpsc.gov/cpsia.pdf [Accessed 10 March 2009].
8.   BS 3084 Slide fasteners (Zips). Specification.
9.   ASTM D 2060 Standard test methods for measuring zipper dimensions.
10.   ASTM D 2061 Standard test methods for strength tests for zippers.
11.   DIN 3419 – 1: Slide fasteners – Part 1: technical delivery conditions.
12.   JIS S 3015 Slide fasteners.

13. CFR Title 16-Part 1500 Consumer product safety commission part 1500 - hazardous substances and articles; administration and enforcement regulations.
14. EN 14682:2004 (E) Safety of children's clothing: cords and drawstrings on children's clothing.

CHAPTER 6

Case studies of safety review in children garment

Abstract

Safety aspects in children's garments are matter of serious concern, and proper care is required to minimize the risk involved in its practical use. The chapter first reviews the safety aspect of different components used in different merchandise items for children with specific case studies. The chapter then discusses the evaluation of restricted substances, such as lead, mercury, antimony, arsenic, barium, cadmium, chromium, selenium, phthalates, and hazardous liquid chemicals. Flammability requirements of children's apparel are also been highlighted.

Keywords: safety, asphyxiation, pull test, toxicity, sleepwear

Chapter contains (Section headings)

6.1 Introduction
6.2 Safety review
    6.2.1 Raglan romper with front pockets (Disney): new born–24 months
    6.2.2 Girls 3 PC raglan hoody, top and pant set (George): 3–24 months
    6.2.3 Boy's 2 PC jersey s/slv top and canvas short pant set (George): 3–24 months
    6.2.4 Girls 3 PC blouse, tank top and capri set (Disney): 3–24 months
    6.2.5 Girls 3 PC jumper dress, top and legging set (Disney): 3–24 months
    6.2.6 Toddler boy short all and tee set (Bum): 2–3X
    6.2.7 3 PC jacket, denim pant and t-shirt set (Disney): 2–3X
    6.2.8 Boys 3 PC shirt, polo top and jean pant set (Disney): 3–24 months
    6.2.9 Key item bottoms–denim (George): 2–3X

6.1   Introduction

An apparel product is considered to be safe, if it does not constitute any risk, or only a very slight risk, to people's health or safety when used under normal or reasonably predictable conditions during its useful life. When making an assessment of whether the risk associated with a product is acceptable and in line with a sufficient level of protection, special consideration is taken of the risks that the product may entail for certain consumer groups, particularly children.

6.2   Safety review

Cords, toggles, and hoods in children's clothing can constitute serious accident risks, and in some cases it can cause death. Some of the common hazards are highlighted to understand the importance of exercising precautions.

Long cords with knots or toggles may get caught when children go on slides. In the event of a cord from a hood gets caught in an opening at the top of the slide, the resultant asphyxiation might be the outcome when the child goes down, and the garment is pulled back. It may also be possible that they may get caught in a bus door. In such a case, if a cord from the lower part of a jacket swings out when the child gets off the bus and gets caught in the doors, the child may become trapped and dragged along when the bus drives on. It is also not unusual phenomenon when they get entangled in a bicycle chain during cycling, if the cords hanging down by the legs.

Apart from the above happenings, asphyxiation can also be caused if the hoods get caught when a child is playing either on a climbing frame or trying to climb a tree. Small parts, such as decorations or buttons can cause choking hazards; if they come out during the use and the child puts them in its mouth. Sharp points and sharp edges in any children's product are no exception to this and any potential health hazard is inevitable.

Restricted substances in children's merchandise, such as lead in surface coating, substrates, and base material or soluble compounds in surface coating, and phthalates in mouthable components are also harmful.

Considering the above facts, it is thought worthwhile to review the safety aspects of different merchandise items for children with specific case studies, which are depicted in this section from Figure 6.1 to Figure 6.16.

6.2.1 Raglan Romper with front pockets (Disney): new born–24 months (Figure 6.1)

1. All garments must be clear of attached or unattached threads to comply with the safety standards.
2. Minimum neck stretch must meet the required measurement specified.
3. Monofilament thread is not permitted for the use on children's clothing.
4. All hardware elements must be sourced from nominated supplier and must comply with the safety standards.
5. All snaps, grommets, rivets must be securely attached to the garment, and must withstand pull test.
6. For all the paints and other coatings used on hardware, including buttons and findings, the lead content must be checked for permissible limit.
7. All buttons must be machine lock-stitched and withstand pull test.
8. All embroidery to be backed with compatible interlining when the reverse of the embroidery is scratchy and comes in direct contact with the skin.[1]
9. If the compatible interlining is utilized, it must be permanently attached; hence, it cannot pose a potential choking hazard.
10. All embroidery paper backing must be completely removed before applying the compatible backing.

Figure 6.1   Raglan Romper with front pockets (Disney): new born–24 months

6.2.2   Girls 3 PC Raglan Hoody, top, and pant set (George): 3–24 months (Figure 6.2)

1. All garments must be clear of attached or unattached threads to comply with the safety standards.
2. Minimum neck stretch must meet the required measurement specified.
3. All hardware elements must be sourced from nominated supplier and must comply with the safety standards.
4. All snaps, grommets, rivets must be securely attached to the garment, and must withstand pull test.
5. Vendors must disclose on sample tag and the nominated sources that will be used in bulk production.
6. All velcro corners must be rounded; velcro tab must withstand pull test.
7. Raw edges on cord ends are not acceptable. Cord ends must be heat sealed, tipped, or clean finished.

6.2.3   Boy's 2 PC Jersey s/slv top and canvas short pant set (George): 3–24 months (Figure 6.3)

1. All garments must be clear of attached or unattached threads to comply with the safety standards.
2. Minimum neck stretch must meet the required measurement specified.
3. Monofilament thread is not permitted for the use on children's clothing.
4. All hardware elements must be sourced from nominated supplier and must comply with the safety standards.

Figure 6.2   Girls 3 pc raglan hoody, top and pant set (George): 3–24 months

5. All snaps, grommets, rivets must be securely attached to the garment, and must withstand pull test.
6. All embroidery to be backed with compatible interlining if the reverse of the embroidery is scratchy and comes in direct contact with the skin.
7. If the compatible interlining is utilized, it must be permanently attached so it cannot pose a potential choking hazard.
8. All embroidery paper backing must be completely removed.

Figure 6.3   Boy's 2 pc jersey top and canvas short pant set (George): 3–24 months

6.2.4   Girls 3 PC blouse, tank top, and capri set (Disney): 3–24 months (Figure 6.4)

1.   All garments must be clear of attached or unattached threads to comply with the safety standards.
2.   Minimum neck stretch must meet the required measurement specified.
3.   Monofilament thread is not permitted for the use on children's clothing.
4.   All hardware elements must be sourced from nominated supplier and must comply with the safety standards.
5.   All snaps, grommets, rivets must be securely attached to the garment, and must withstand pull test.
6.   For all the paints and other coatings used on hardware including buttons and findings, the lead content must be checked for permissible limit.
7.   All buttons must be machine lock-stitched and withstand pull test.
8.   Belt loop must be a maximum of 1½″ in length between bartacks on size 0–3x.
9.   All embroidery to be backed with compatible interlining if the reverse of the embroidery is scratchy and comes in direct contact with the skin.
10.   If the compatible interlining is utilized, it must be permanently attached so it cannot pose a potential choking hazard.
11.   All embroidery paper backing must be completely removed.
12.   Graphic label/patch must be securely attached to withstand pull test.

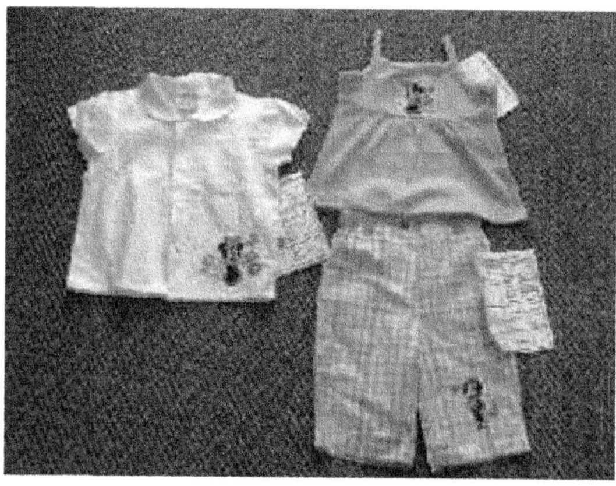

Figure 6.4   Girls 3 PC blouse, tank top, and capri set (Disney): 3–24 months

6.2.5   Girls 3 PC jumper dress, top and legging set (Disney): 3–24 months (Figure 6.5)

1. All garments must be clear of attached or unattached threads to comply with the safety standards.
2. Minimum neck stretch must meet the required measurement specified.
3. Correct components are to be used that comply with the safety standards.
4. Vendors must disclose on tag and the nominated sources that will be used in bulk production.
5. All hardware elements must be sourced from nominated supplier and must comply with the safety standards.
6. All snaps, grommets, rivets must be securely attached to the garment, and must withstand pull test.
7. All embroidery to be backed with compatible interlining if the reverse of the embroidery is scratchy and comes in direct contact with the skin.
8. If the compatible interlining is utilized, it must be permanently attached; hence, it cannot pose a potential choking hazard.
9. All embroidery paper backing must be completely removed.
10. All embroidery paper backing must be completely removed before applying the compatible backing.
11. Thread floats and thread ends over 3/8″ are not acceptable on children's clothing.
12. Graphic label/patch must be securely attached to withstand the pull test.

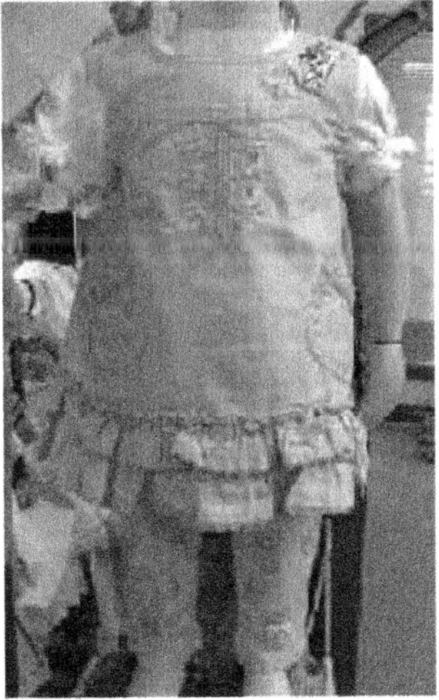

Figure 6.5   Girls 3 pc jumper dress, top and legging set (Disney): 3–24 months

6.2.6 Toddler boy short all and tee set (Bum): 2–3X (Figure 6.6)

1. Minimum neck stretch must meet the required measurement specified.
2. Monofilament thread is not permitted for the use on children's clothing.
3. All hardware elements must be sourced from nominated supplier and must comply with the safety standards.
4. All snaps, grommets, rivets must be securely attached to the garment, and must withstand pull test.
5. Correct components are to be used those comply with the safety standards.
6. Vendors must disclose on tag and the nominated sources that will be used in bulk production.

6.2.7 3 PC jacket, denim pant, and t-shirt set (Disney): 2–3X (Figure 6.7)

1. Garment must be free of any attached or unattached thread ends to comply with the safety standards.
2. Minimum neck stretch must meet the required measurement specified.
3. All snaps, grommets, rivets must be securely attached to the garment, and must withstand pull test.
4. Belt loop must be a maximum of 1½″ in length between bartacks on size 0–3x.
5. Zippers must be sourced from approved suppliers only.
6. Zippers must have fully autolock or semi-autolock sliders.

Figure 6.6 Toddler boy short all and tee set (Bum): 2–3X

7. Molded plastic zipper top stops must have a hook or ball to prevent zip slider from detaching while in open state.
8. All hardware elements must be sourced only from nominated supplier and comply with the safety standards.
9. The components cannot have rough or sharp edges; and exposed prongs are prohibited.
10. All components should be free of oxidation, rust, or other types of degraded corrosion.
11. For all the paints and other coatings used on hardware, including buttons and findings, the lead content must be tested for permissible limit as per CPSIA 2008.[2]
12. All embroidery to be backed with compatible interlining, if the reverse of the embroidery is scratchy and comes in direct contact with the skin.
13. If the compatible interlining is utilized, it must be permanently attached; therefore, it cannot pose a potential choking hazard.
14. All embroidery paper backing must be completely removed.
15. Thread floats and thread ends over 3/8″ are not acceptable on children's clothing.

6.2.8 Boys 3 PC shirt, polo top and jean pant set (Disney): 3–24 months (Figure 6.8)

1. All garments must be clear of attached or unattached threads to comply with the safety standards.
2. Minimum neck stretch must meet the required measurement specified.
3. Monofilament thread is not permitted for the use on children's clothing.

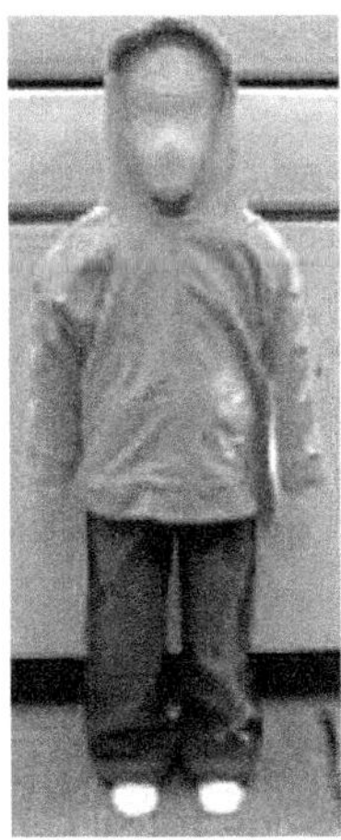

Figure 6.7   3 PC jacket, Denim pant, and t-shirt set (Disney): 2–3X

4. For all the paints and other coatings used on hardware, including buttons and findings, the lead content must be tested for permissible limit.
5. All buttons must be machine lock-stitched and withstand pull test.
6. All velcro corners must be rounded; and velcro tab must withstand pull test.
7. Belt loop must be a maximum of 1½″ in length between bartacks on size 0–3x.
8. Convertible tabs must not be exceeding 1½″ in length from point of secure attachment for size 0-3x.
9. All side tab loop labels must not exceed ½″ (folded) from surface of fabric or stitch down both side to prevent the loop forming.
10. All embroidery to be backed with compatible interlining if the reverse of the embroidery is scratchy and comes in direct contact with the skin.
11. If the compatible interlining is utilized, it must be permanently attached; hence, it cannot pose a potential choking hazard.
12. All embroidery paper backing must be completely removed.
13. Graphic label/patch must be securely attached to withstand pull test.
14. All pocket tabs must not exceed 1½″ (folded) from surface of fabric and must bartack at centre of tabs.

6.2.9 Key item bottoms–Denim (George): 2–3X (Figure 6.9)

1. Garment must be free of any attached or unattached thread ends to comply with the safety standards.
2. All buttons must be machine lock-stitched and withstand pull test.
3. Belt loop must be a maximum of 1½″ in length between bartacks on size 0–3x.
4. Adjustable elastic must be securely tacked at centre back waistband.

Figure 6.8 Boys 3 PC shirt, polo top, and jean pant set (Disney): 3–24 months

5. All hardware elements must be sourced only from nominated supplier and comply with the safety standards.
6. The components cannot have rough or sharp edges; exposed prongs are prohibited.
7. All components should be free of oxidation, rust, or other types of degraded corrosion.
8. All snaps, must be securely attached to the garment, and must withstand pull test.
9. For all the paints and other coatings used on hardware, including buttons and findings, the lead content must be tested for permissible limit.
10. All metal fastenings and metal buttons must comply with CPSIA 2008 regulation.
11. Zippers must be sourced from approved suppliers only.
12. Zippers must have fully autolock or semi-autolock sliders.
13. Metal zipper required a zipper guard or facing to prevent the zipper being in direct contact with the skin.
14. Detachable fabric belt is allowed; tail ends must not longer than 3″ in double tied state on 0–3x.
15. Tail ends of sash must be clean finished.
16. Convertible tabs must be no exceed 1½″ in length from point of secure attachment for 0–3x.

Figure 6.9   Key item bottoms–Denim (George): 2–3X

17. Vendors must disclose on hangtag and the nominated sources that are used in bulk production.
18. All embroidery must be backed with compatible interlining if the reverse of the embroidery is scratchy and comes in direct contact with skin.
19. If the compatible interlining is utilized, it must be permanently attached; hence, it cannot pose a potential choking hazard.
20. All embroidery paper backing must be completely removed.

6.2.10 Girls 3 PC jersey top, sweater card, and Denim pant set (George): 3–24 months (Figure 6.10)

Safety review

1. All garments must be clear of attached or unattached threads to comply with the safety standards.
2. Minimum neck stretch must meet the required measurement specified.
3. Monofilament thread is not permitted for the use on children's clothing.
4. Preproduction sample will be subject to fail if incorrect components are used and do not comply with the safety standards.
5. Vendors must disclose on sample tag and the nominated sources that will be used in bulk production.
6. All hardware elements must be sourced from nominated supplier and must comply with the safety standards.
7. All snaps, grommets, rivets must be securely attached to the garment, and must withstand pull test.
8. For all the paints and other coatings used on hardware, including buttons and findings, the lead content must be checked for permissible limit.
9. Buttons on sweater knit are accepted by hand sewn finishing as long as pass the pull test.
10. All buttons must be machine lock-stitched and withstand pull test.
11. Belt loop must be a maximum of 1½" in length between bartacks on 0–3x.
12. All embroidery paper backing must be completely removed.
13. All embroidery paper backing must be completely removed before applying the compatible backing.
14. Thread floats and thread ends over 3/8" are not acceptable on children's clothing.

6.2.11   Girls romper (Disney): 3–24 months (Figure 6.11)

Safety review

1. All garments must be clear of attached or unattached threads to comply with the safety standards.
2. Minimum neck stretch must meet the required measurements specified.
3. All hardware elements must be sourced only from nominated supplier and comply with the safety standards.
4. All snaps, grommets, rivets must be securely attached to garment, and must withstand pull test.
5. For all the paints and other coatings used on hardware, including buttons and findings, the lead must be evaluated to check permissible limits as per CPSIA 2008 is essential.
6. All buttons must be machine lock-stitched and withstand pull test.
7. All embroidery to be backed with compatible interlining if the reverse of the embroidery is scratchy and comes in direct contact with the skin.
8. If the compatible interlining is utilized, it must be permanently attached; hence, it cannot pose a potential choking hazard.
9. All embroidery paper backing must be completely removed.
10. Thread floats and thread ends over 3/8″ are not acceptable on children's clothing.

Figure 6.10   Girls 3 PC jersey top, sweater card, and Denim pant set (George): 3–24 months

6.2.12 Boys Raglan Romper (Disney): new born–24 months (Figure 6.12)

Safety review

1. All garments must be clear of attached or unattached threads to comply with the safety standards.
2. Monofilament thread is noZt permitted for the use on children's clothing.
3. Correct components are to be used to comply with the safety standards.
4. Components from non-nominated suppliers are to be used.
5. Vendors must disclose on sample tag and the nominated sources that will be used in bulk production.
6. All hardware elements must be sourced from nominated supplier and must comply with the safety standards.
7. All snaps, grommets, rivets must be securely attached to the garment, and must withstand pull test.
8. 3D ears measured 1¼″ height and 1¼″ width, and must be securely attached to withstand pull test.
9. All embroidery to be backed with compatible interlining if the reverse of the embroidery is scratchy and comes in direct contact with the skin.
10. If the compatible interlining is utilized, it must be permanently attached; hence, it cannot pose a potential choking hazard.
11. All embroidery paper backing must be completely removed.

Figure 6.11 Girls Romper (Disney): 3–24 months3X

## 6.2.13	Girls Raglan Romper (Disney): new born–24 months (Figure 6.13)

Safety review

1.	All garments must be clear of attached or unattached threads to comply with the safety standards.
2.	Monofilament thread is not permitted for the use on children's clothing.
3.	Correct components are to be used to comply with the safety standards.
4.	Components shall be from nominated suppliers.
5.	Vendors must disclose on tag and the nominated sources that will be used in bulk production.
6.	All hardware elements must be sourced from nominated supplier and must comply with the safety standards.
7.	All snaps, grommets, rivets must be securely attached to the garment, and must withstand pull test.
8.	3D ears measured 3″ and 1 1/2″ from fabric surface and must be securely attached to withstand pull test.
9.	All embroidery to be backed with the compatible interlining if the reverse of the embroidery is scratchy and comes in direct contact with the skin.

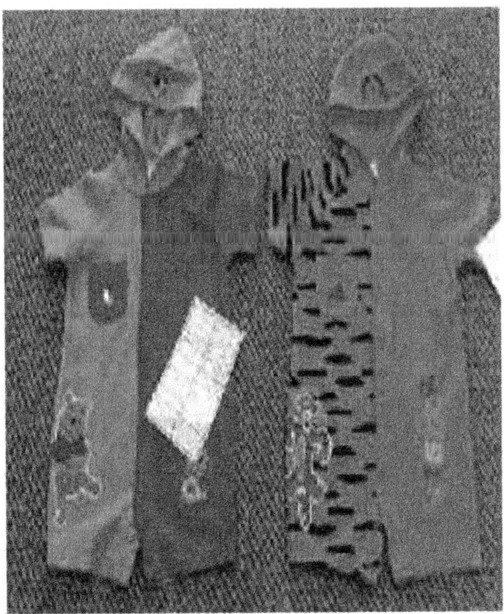

Figure 6.12	Boys Raglan Romper (Disney): new born–24 months

10. If the compatible interlining is utilized, it must be permanently attached; hence, it cannot pose a potential choking hazard.
11. All embroidery paper backing must be completely removed.

6.2.14 Boys and girls colour blocked applique sleeper (Disney): 3–24 months (Figure 6.14)

Safety issues

1. All garments must be clear of attached or unattached threads to comply with the safety standards.
2. Children's sleepwear must meet the required dimensions specified from health Canada guidelines for polo pajamas and sleepers, loose edges up to 5 cm (2") are permitted at the neck only. Sample complies which measures: 3/4" wide neckband.
3. Only flammability testing can ensure that a fabric complies with the children's sleepwear regulations.
4. All hardware elements must be sourced from nominated supplier and must comply with the safety standards.
5. All snaps, grommets, rivets must be securely attached to the garment, and must withstand pull test.
6. All embroidery to be backed with the compatible interlining if the reverse of the embroidery is scratchy and comes in direct contact with the skin.
7. If the compatible interlining is utilized, it must be permanently attached; hence, it cannot pose a potential choking hazard.
8. All embroidery paper backing must be completely removed.
9. Graphic label/patch must be securely attached to withstand the pull test.
10. The presence of cotton thread, trims, or decoration on a 100% nylon, 100% polyester, or polyester/nylon blends may affect the flammability of the garment.

6.2.15 Toddler girls woven short (George): 2–3X (Figure 6.15)

Safety review

1. Garment must be free of any attached or unattached thread ends to comply with the safety standards.
2. All buttons must be machine lock-stitched and withstand pull test.
3. Adjustable elastic must be securely tacked at the centre back of the waistband.
4. All hardware elements must be sourced only from nominated supplier and comply with the safety standards.

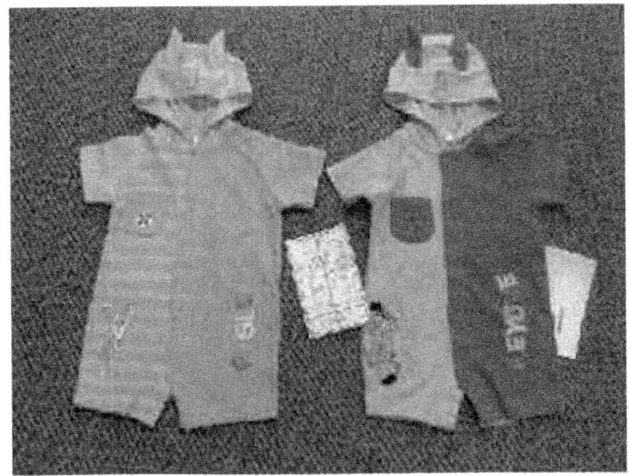

Figure 6.13   Girls Raglan Romper (Disney): new born–24 monthsmonths3X

5. The components cannot have rough or sharp edges; and exposed prongs are prohibited.
6. All components should be free of oxidation, rust, or any other types of degraded corrosion.
7. All snaps must be securely attached to the garment and must withstand pull test.
8. For all the paints and other coatings used on hardware, including buttons and findings, the lead content must be tested for permissible limit.
9. All metal fastenings and metal buttons must comply with CPSIA 2008 regulation.
10. Zippers must be sourced from the approved suppliers only.
11. Zippers must have fully autolock or semi-autolock sliders.
12. Metal zipper required a zipper guard or facing to prevent the zipper being in direct contact with the skin.
13. Vendors must disclose on hangtag and the nominated sources that are used in bulk production.

6.2.16   2 PC Hawaiin dress set: Disney (3–24 months) (Figure 6.16)

Safety review

1. Minimum neck stretch must meet the required measurement specified. Neck opening when extended measures 21″.
2. Monofilament thread is not permitted for the use on children's clothing.
3. Fixed bows/rosettes must be securely tacked at centre and must withstand pull test.

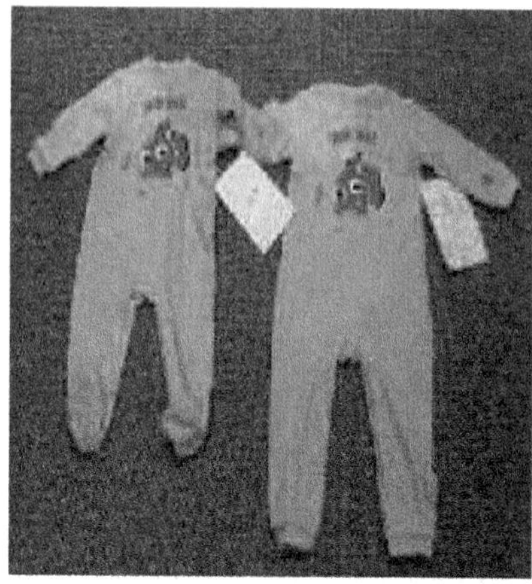

Figure 6.14   Boys and girls colour blocked applique sleeper (Disney): 3–24 months

4.   The loop and tail lengths cannot exceed the following: bow loops (folded) 3/4″; tail length 1″. Sample measures: loop ½″ (folded) and tail length ¾″.

5.   Raw edges on cord ends are not acceptable. Cord ends must be heat sealed, tipped, or clean finished.

6.   All side tab loop labels must not exceed 1″ (folded) from surface of fabric or stitch down both side to prevent the loop forming. Tab loop on sample measures 3/8″ flat; suggest stitching down both sides to prevent loop from forming.

6.3   Evaluation of restricted substances and hazardous components testing for children's products

Lead is a cumulative toxic heavy metal, which presents a chronic health hazard, especially in children. The effects of lead poisoning include hyperactivity, slowed learning ability, withdrawal, blindness, and even death. Thus, total lead in surface coatings and substrate materials are restricted in children's products. Other substances, such as mercury, antimony, arsenic, barium, cadmium, chromium, selenium, phthalate, and hazardous liquid chemicals have also adverse effects on children's health.

6.3.1   Total lead in surface coatings

Product should meet the following condition:

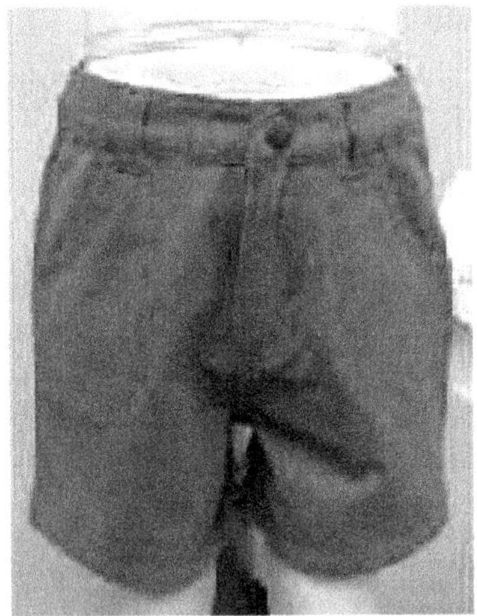

Figure 6.15   Toddler girls woven short (George): 2–3X

The surface coating(s) scraped from the product do not exceed 0.009% (90ppm). However, before July 1, 2009, a surface coating lead level of 0.06% (600 ppm) is acceptable for in-store products.

6.3.2   Total lead in substrates/base materials

Product should meet the following condition:

All accessible substrate/base materials of the finished goods do not exceed 0.03% (300 ppm). However, before July 1, 2009, a substrate/base material lead level of 0.06% (600 ppm) is acceptable for in-store products.

6.3.3   Soluble compounds/elements in surface coatings

The following compounds in surface coating(s) scraped from the product do not exceed the following limits [ASTM F963 Section 4.3.5.2[3]]:

Lead	0.009% (90 ppm) soluble
Mercury	not detectable
Antimony	0.006% (60 ppm) soluble
Arsenic	0.0025% (25 ppm) soluble
Barium	0.1% (1000 ppm) soluble
Cadmium	0.0075% (75 ppm) soluble
Chromium	0.006% (60 ppm) soluble
Selenium	0.05% (500 ppm) soluble

6.3.4   Hazardous liquid chemicals:

No hazardous chemicals should be present in liquid filled products in accordance with 16 CFR 1500.231.[4]

6.3.5   Toxicity and irritancy

Toxicological risk assessment report (TRA) should indicate that all liquids, putties, pastes, powders, and gels have been assessed for toxicity risk.

USP 61 microbial limits test: Evaluation should indicate that liquids, putties, pastes, powders, or gels sampled from the finished goods have been tested in accordance to United States Pharmacopoeia (USP) 61 microbial limits test[5] and are within the following limits:

Tested age	Total viable count limit
18 months or less	500 cfu/ml (g)
Over 18 months	5000 cfu/ml (g)

The absence of Staphylococcus aureus, Pseudomonas aeruginosa, Salmonella, and Escherichia coli should be confirmed in the evaluation.

USP 51 preservative effectiveness test: Evaluation of liquids, putties, pastes, powders, or gels sampled from the finished goods should be done in accordance to United States Pharmacopeia (USP) 51 preservative effectiveness test.[6]

Combustible liquids: All liquids sampled from the finished goods should have a flash point higher than 150°F (65.6°C) when tested in accordance with 16 CFR 1500.43.[7]

Figure 6.16   16.   2 PC Hawaiin dress set: Disney (3–24 months)

6.3.6 Phthalates

The finished product composed of (or has components composed of) accessible soft and pliable PVC, or plasticized materials contains no more than 0.1% (1000 ppm) of the following phthalates. Evaluation can be performed as per method EPA 8270C[8] or EN 14372.[9]

Child care articles for children under 4, and all mouthable accessible components of toys:

DINP (Diisononyl phthalate
DIDP (Diisodecyl phthalate)
DnOP (Di-*n*-octyl phthalate)
DEHP (Di-(2-ethylhexyl) phthalate)
BBP (Benzyl butyl phthalate)
DBP (Dibutyl phthalate)
For accessible components of all other toys and child care items:
DEHP (Di-(2-ethylhexyl) phthalate)
BBP(Benzyl butyl phthalate)
DBP(Dibutyl phthalate)

6.3.7 Mechanical Hazards

The finished item does not present any potential hazard for children (sharp edges, sharp points, small parts) as required in 16 CFR 1500[10] and 1501.[11]

6.4 General wearing apparel flammability requirements

All children's wearing apparel must comply with any and all state and/ or federal guidelines, regulations, and laws, including but not limited to reasonable and representative testing as required under CFR Title 16, Part 1610, and must be classified as Class 1: normal flammability.

6.5 Children's sleepwear/loungewear flammability requirements

All children's sleepwear sized 0–18 must comply with any and all state and/ or federal guidelines, regulations, and laws, including but not limited to CFR Title 16, Part 1615, and 1616. If the packaging or any printed design on a children's garment suggests sleeping, it will be considered sleepwear, and must pass testing to comply with the requirements set forth in CFR Title 16, Part 1615, and 1616. Loungewear is considered sleepwear, and; therefore, must follow the same requirements as sleepwear. CPSC is the final authority for determining what loungewear is.

Federal flammability requirements for children's sleepwear are outlined in chapter 4 of this book. Even though the sleepwear for infants under 9M is exempt from the children's sleepwear regulations, the sleepwear must still meet the general wearing apparel flammability requirements. Children's sleepwear sized above 9M and upto 6X must meet the flammability requirements of 16 CFR 1615, whereas and the flammability of children's sleepwear sizes 7 through 14 is covered under 16 CFR 1616. Although tight fitting garments are exempt from the flammability requirements of 16 CFR 1615 and 1616, they must meet the sizing and labeling requirements specified in 16 CFR 1615 and 1616 and meet Class I normal flammability under 16 CFR 1610 for general wearing apparel.

References

1. Embroidery technologies, LLC, Introduction to embroidery backings. Available from: http://www.embroiderytechnologies.com/Backinginfo.pdf [Accessed 9 March, 2009].

2. Consumer product safety improvement act of 2008, Public law 110-314. Available from: http://www.cpsc.gov/cpsia.pdf [Accessed 13 March, 2009].

3. ASTM F963 Section 4.3.5.2 Standard consumer safety specification on toy safety.

4. 16 CFR 1500.231 Guidance for hazardous liquid chemicals in children's products.

5. Swarbrick James and Boylan C James (2001), USP chapter (61) antimicrobial limits tests, Current microbiological testing practices, *Encyclopedia of Pharmaceutical Technology*, Published by Marcel Dekker, New York, 20, 222.

6. Swarbrick James and Boylan C James (2001), USP chapter (51) antimicrobial effectiveness test, Current microbiological testing practices, *Encyclopedia of Pharmaceutical Technology*, Published by Marcel Dekker, New York, 20, 222.

7. 16 CFR 1500.43 Method of test for flashpoint of volatile flammable materials by tagliabue open-cup apparatus.

8. EPA 8270C Semivolatile organic compounds by Gas chromatography/Mass spectrometry (GC/MS).

9. EN 14372 Child use and care articles. Cutlery and feeding utensils. Safety requirements and tests.

10. 16 CFR 1500 Consumer product safety commission part 1500 - hazardous substances and articles; administration and enforcement regulations.

11. 16 CFR Part 1501 Method for Identifying toys and other articles intended for use by children under 3 years of age which present choking, aspiration, or ingestion hazards because of small parts.

CHAPTER 7

Product recall in children garment

Abstract

Product recall entails returning of goods to the selling point for a full return of payment or modification due to defective product or over safety issues. The chapter first discusses the causes of product recall to protect the consumer and to retain the image of a brand. The chapter then discusses various aspects of product recall for children's apparel and related items with the help of different case studies.

Keywords: recall, defects, protection, strangulation, choking

Chapter contents (Section headings)

7.1 Reason of product recall
7.2 Case studies on product recall
  7.2.1 Boy's jackets
  7.2.2 Children's pants
  7.2.3 Girl's clothing sets
  7.2.4 Children's sweaters
  7.2.5 Girls' hooded sweatshirts with drawstrings
  7.2.6 Children's hooded sweatshirts
  7.2.7 Children's hooded sweatshirts
  7.2.8 Pajamas
  7.2.9 Sleeping bags
  7.2.10 Infant garment
  7.2.11 Children's board skirts
  7.2.12 Children's bobbie socks
  7.2.13 Hooded sweaters
  7.2.14 Heat transferred, or "Tag-less," labels
  7.2.15 "Feather witch" Halloween costume (Canada)
  7.2.16 N-Kids brand drawstring flannel pants
  7.2.17 Newborn and infant pants

7.1 Reason of product recall

A product recall is a request to return to the maker, a batch or an entire production run of a product, usually over safety concerns or design defects or labeling errors.[1] Generally speaking, companies recall products when defects seem to have safety concerns for customers and affect large number of customers. Otherwise normal warranty procedures are adequate as a part of standard marketing practice. Thus, recalls are more of preventive in nature and to pre-empt costly litigations and financial and goodwill losses. Product recalls are governed by consumer protection laws[2] of a country that have specific requirements that cover the extent of cost the manufacturer will have to bear in which a recall is compulsory and penalties in case of failure to recall. In the event of a product recall, announcements may be released on the respective government agency's website; notices can be given in the metropolitan daily newspapers and also in certain circumstances recall may be advised in news television reports as heightened publicity. During such recall, the consumer is requested to return goods, irrespective of condition, to the selling point for a full refund or modification.

7.2 Case studies on product recall

Different actual cases of product recall in Children's Garment are highlighted which has been happened in recent years.

7.2.1 Boy's jackets (Figure 7.1)

Jackets have a waist drawstring with a toggle that could become snagged or caught in small spaces or doorways, which can pose an entrapment hazard[3] to children. In February 1996, CPSC issued guidelines to help prevent children

Figure 7.1   Boy's jackets

from getting entangled on waist by drawstrings in upper garments,[4] such as jackets and sweatshirts.
Store responsible for the product: Old Navy
Month of recall: December 2007
Country of origin: Indonesia

7.2.2   Children's pants (Figure 7.2)

Recalled pants have a ribbon belt at the waist that can pose an entrapment or entanglement hazard.[5]
Store responsible for the product: Sears stores
Month of recall: December 2007
Country of origin: China

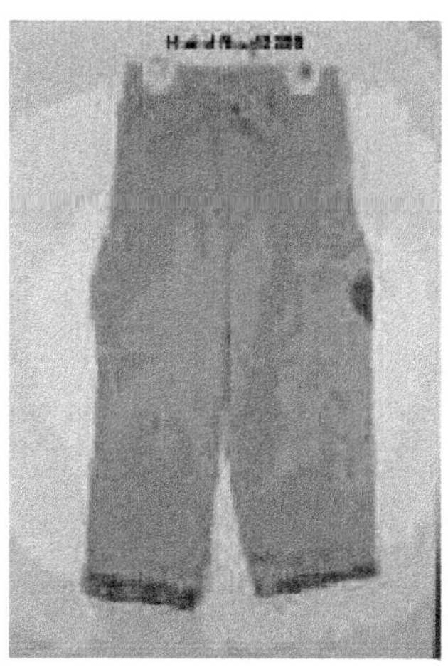

Figure 7.2   Children's pants

7.2.3   Girl's clothing sets (Figure 7.3)

Recalled pants have a drawstring at the waist that can pose an entrapment or entanglement hazard to children.[6]
Store responsible for the product: K-Mart stores
Month of recall: December 2007
Country of origin: Pakistan

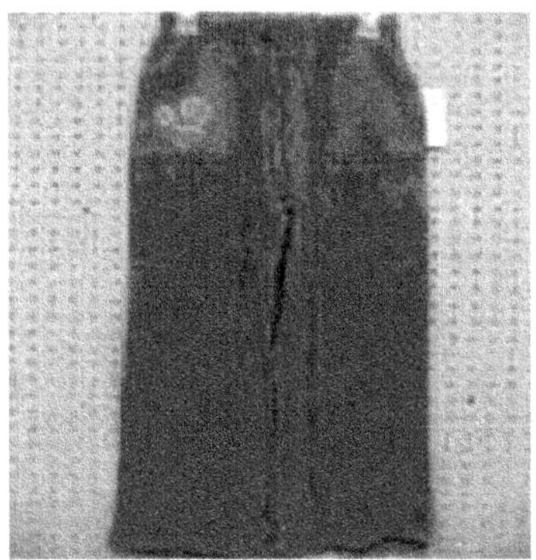

Figure 7.3   Girl's clothing sets

7.2.4   Children's sweaters (Figure 7.4)

Recalled sweaters have a drawstring through the hood, posing a strangulation hazard[7] to children.
Store responsible for the product: Sears stores
Month of recall: December 2007
Country of origin: Pakistan

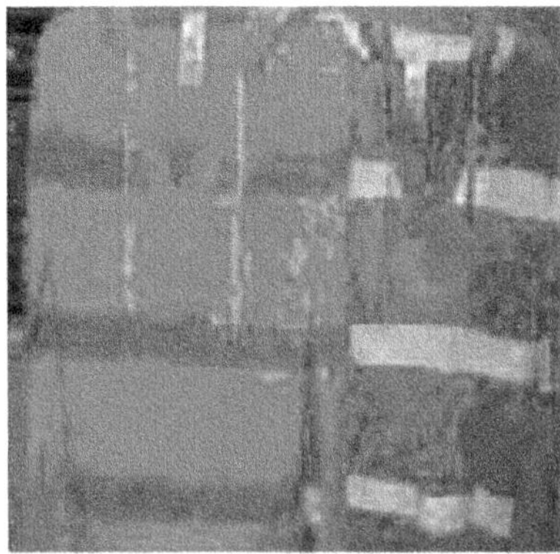

Figure 7.4   Children's sweaters

7.2.5   Girls' hooded sweatshirts with drawstrings (Figure 7.5)

Garments have a drawstring through the hood, which can pose a strangulation hazard to children.[8]
Store responsible for the product: Marshalls and other specialty children's clothing retailers nationwide
Month of recall: December 2007
Country of origin: India

Figure 7.5   Girls' hooded sweatshirts with drawstrings

7.2.6   Children's hooded sweatshirts (Figure 7.6)

Garments have a drawstring through the hood, which can pose a strangulation hazard to children.[9]
Store responsible for the product: Nordstrom stores
Month of recall: March 2008
Country of origin: China

Figure 7.6   Children's hooded sweatshirts

7.2.7   Children's hooded sweatshirts (Figure 7.7)

Garments have a drawstring through the hood, which can pose a strangulation hazard to children.[10]
Store responsible for the product: Various retail stores nationwide
Month of recall: April 2008
Country of origin: China

Figure 7.7   Children's hooded sweatshirts

7.2.8   Pajamas (Figure 7.8)

The sleeve opening to the pajama top is too large and does not conform to Canadian Flammability regulations.[11] If a child is too close to a flame, the sleeve could catch fire.
Store responsible for the product: Sears Canada retail outlets.
Month of recall: May 2008
Country of origin: India

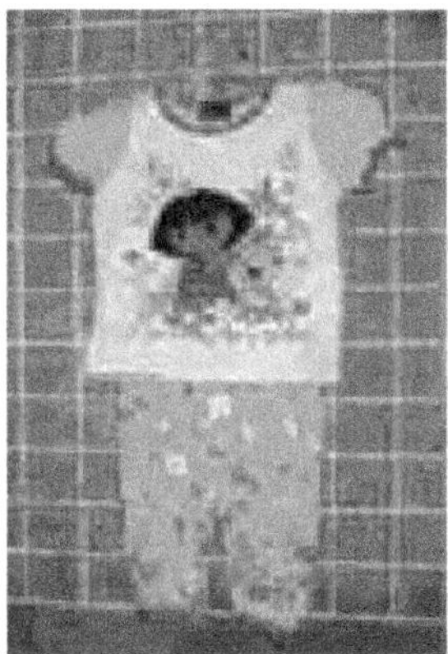

Figure 7.8   Pajamas

7.2.9 Sleeping bags (Figure 7.9)

Surface paint on the sleeping bag's zipper pull contains excessive levels of lead, violating the federal lead paint standard.[12]
Store responsible for the product: Disney stores
Month of recall: May 2008
Country of origin: China

Figure 7.9 Sleeping bags

7.2.10 Infant garment (Figure 7.10)

Snaps on these garments can detach, posing a choking hazard[13] to young children
Stores responsible for the product: Dillards, Nordstrom, and other specialty stores and Internet retailers
Month of recall: July 2008
Country of origin: China

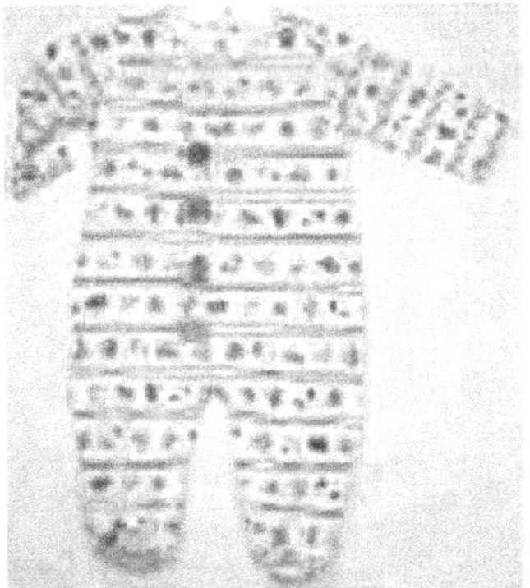

Figure 7.10   Infant garment

7.2.11   Children's board skirts (Figure 7.11)

Paint on the grommets of the skirts contains an excess level of lead,[14] violating the federal lead paint standard.
Store responsible for the product: Chelsea & Scott Ltd.
Month of recall: September 2008
Country of origin: China

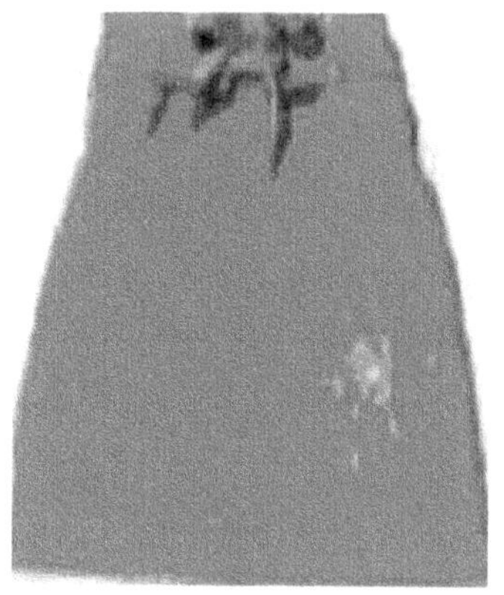

Figure 7.11   Children's board skirts

7.2.12   Children's bobbie socks (Figure 7.12)

Ribbon on the sock can detach[15] posing a choking hazard to young children.
Store responsible for the product: Target stores
Month of recall: September 2008
Country of origin: Hong Kong

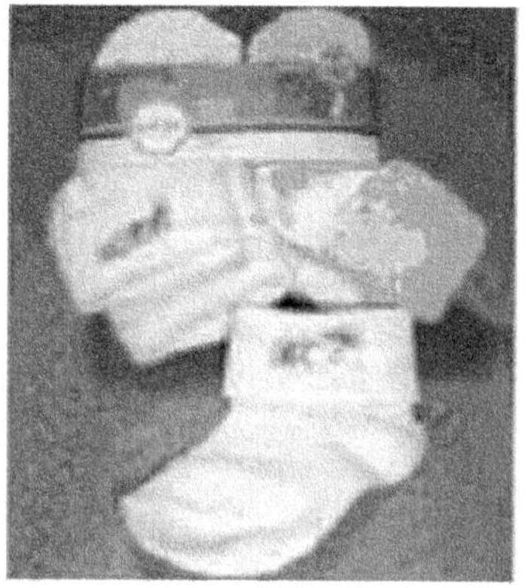

Figure 7.12   Children's bobbie socks

7.2.13   Hooded sweaters (Figure 7.13)

The sweaters have drawstrings through the hood[16]. Children can get entangled in the drawstrings that can catch on playground equipment, fences, or tree branches.
Store responsible for the product: Specialty children's stores
Month of recall: September 2008
Country of origin: China

Figure 7.13   Hooded sweaters

7.2.14   Heat transferred, or "Tag-less," labels (Figure 7.14)

A small percentage of babies and infants have developed rashes on the upper back after wearing Carter's clothing with heat transferred, or "tagless" labels.[17]
Store responsible for the product: Carter's retail stores and their departmental and chain stores
Month of recall: October 2008
Country of origin: Various countries

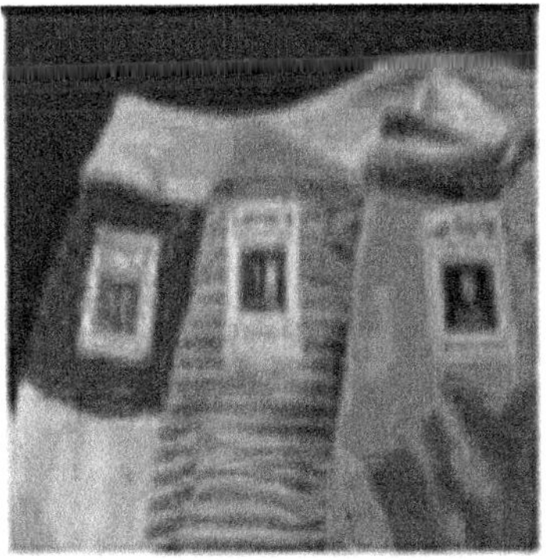

Figure 7.14   Heat transferred, or "Tag-less," labels

7.2.15 "Feather witch" Halloween costume (Canada) [Figure 7.15]

Feathers on the Halloween costumes[18] do not meet the requirements for textile flammability under Canadian law.
Stores responsible for the product: Various retail and novelty stores in Ontario and Quebec
Month of recall: October 2008
Country of origin: China

Figure 7.15   "Feather witch" halloween costume (Canada)

7.2.16 N-Kids brand drawstring flannel pants [Figure 7.16]

These lounge pants are 100 percent cotton and fail to meet the children's sleepwear flammability standards, posing a risk of burn injury to children. These garments[19] were not labeled or marketed as sleepwear, but because they are children's loungewear, they must meet the children's sleepwear flammability standards.
Stores responsible for the product: Nordstrom
Month of recall: March 2007
Country of origin: India

Figure 7.16   Pacifier clip (Canada)

7.2.17   Newborn and infant pants (Figure 7.17)

Metal snap at the waist can detach posing a choking hazard to infants.[20]
Store responsible for the product: J C Penney Co.
Month of recall: November 2008
Country of origin: Bangladesh

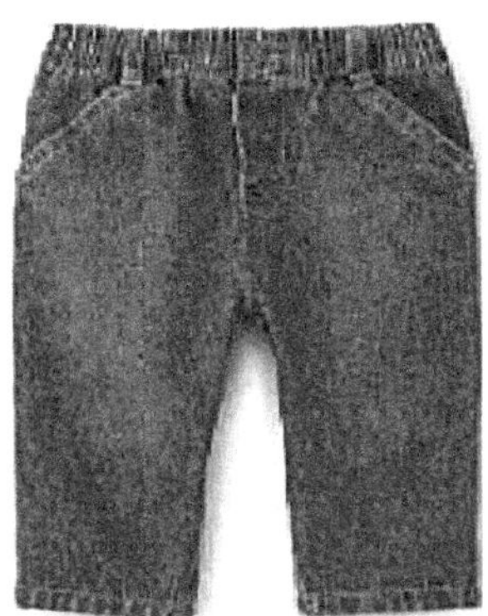

Figure 7.17   Newborn and infant pants

7.2.18   Doll clothing sets (Figure 7.18)

Surface paints on the pajama pants[21] contain excessive levels of lead, which violates the federal lead paint standard.
Store responsible for the product: Manhattan Group
Month of recall: December 2008
Country of origin: Indonesia

Figure 7.18   Doll clothing sets

7.2.19   Toddler girl's hat and mitten sets (Figure 7.19)

Magnets in the hat can detach and fall out, posing a choking and aspiration hazard to young children.[22] Magnets found by young children can be swallowed or aspirated. If more than one magnet is swallowed, the magnets can attract each other and cause intestinal perforations or blockages, which can be fatal. Stores responsible for the product: Meijer stores in Michigan, Illinois, Indiana, Ohio, and Kentucky
Month of recall: December 2008
Country of origin: China

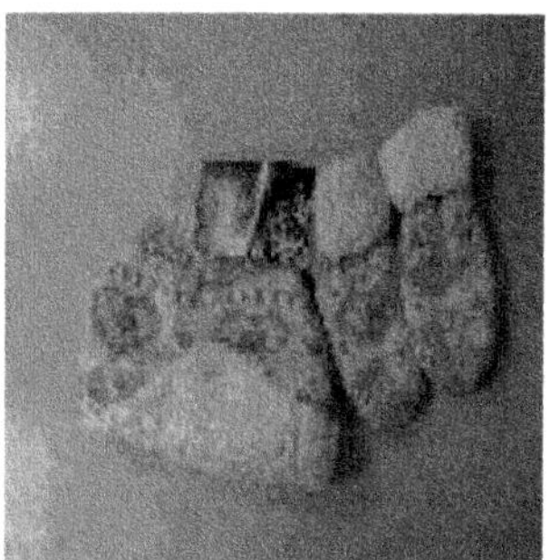

Figure 7.19   Toddler girl's hat and mitten sets

7.2.20   Children's hooded jackets [Figure 7.20]

The drawstrings through the hood and at the waist can catch on playground equipment, fences, or tree branches, posing entanglement and strangulation hazards for children.[23]
Store responsible for the product: R&D International Inc.
Month of recall: November 2008
Country of origin: Indonesia and Peru

Figure 20   Children's Hooded Jackets (Canada)

7.2.21 Precious Cargo infant one-piece garments (Figure 7.21)

Precious Cargo infant one-piece garments with a three snap bottom closure. The garments were 100 percent cotton and sold in sizes 6M, 12M, and 18M in the following solid colors: athletic heather, aquatic blue, candy pink, clover green, jet black, lime, navy, purple, red, royal, sangria, white, and yellow. "Precious Cargo" and elephant are printed on a blue tag sewn into the neck area.[24]

The snaps on the one-piece garments detach posing a choking hazard to young children.

Stores responsible for the product: Promotional product distributors, screen printers, embroiderers, and gift shops

Importer(s): SanMar Corp. of Issaquah, Wash.

Month of recall: August 6, 2015

Country of origin: Vietnam

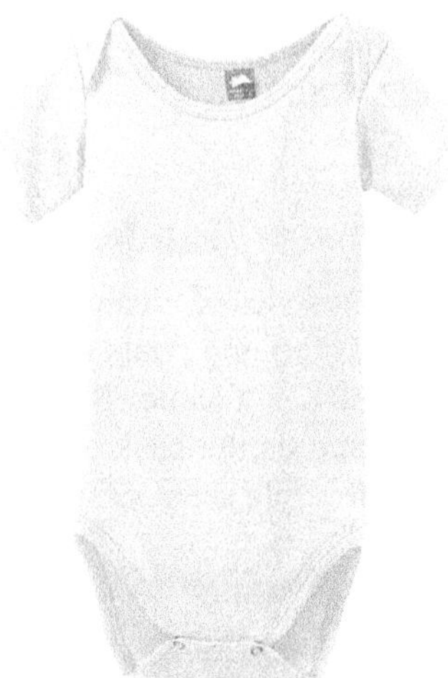

Figure 7.21 Precious Cargo infant one-piece garments

7.2.22 Girl's Three-Piece clothing sets (Figure 7.22)

This recall involves girl's "Young Hearts" brand three-piece clothing sets. The sets were sold with a pink vest, black pullover shirt, and knit pants in sizes 12 months to 6X. "Young Hearts" is printed on a label inside the shirt collar. The pink vest has a black bow applique on the left front and a pink elastic belt with silver clasps.[25]

The vest sold with these sets has a faux half-belt sewn into the side seams with a hook and eye closure at the waist that could become snagged or caught in small spaces or vehicle doors and it poses an entanglement hazard.

Figure 7.22   Girl's Three-Piece Clothing Sets

Stores responsible for the product: Conway, Citi Trends, Duckwall-Alco, and other children's apparel stores nationwide, and online at Amazon.com.
Importer: Children's Apparel Network, of New York, NY.
Month of recall: December 24, 2013
Country of origin: China

7.2.23   Children's loungewear (Figure 7.23)

This recall involves two different styles of Eleanor Rose loungewear, including a girl's gown and a boy or girl's top and pants set. The loungewear was sold in sizes 12 months through size 12. "Eleanor Rose" is printed on a tag sewn into the neck of the garments and on the back outside of the pants. The style number is on a tag sewn into the side seam or inside the back of the pants.[26]

Figure 7.23   Children's loungewear

The loungewear fails to meet federal flammability standards for children's sleepwear posing a risk of burn injuries to children.
Stores responsible for the product: Online at www.eleanorrose.com
Importer: Eleanor Rose, of Natchez, Miss.
Month of recall: April 5, 2016
Country of origin: El Salvador

7.2.24   Children's knit denim jackets (Figure 7.24)

This recall involves Tea Collection Inc. children's knit blue denim jackets with metal buttons and snaps. A tag sewn inside the neck reads "Tea." Style number 6F22400-405 is printed on a hangtag attached to the garment.[27]

The metal snaps on the jackets can detach, posing a choking hazard to children.

Stores responsible for the product: Specialty and other stores nationwide.
Distributor: Tea Living Inc., d/b/a Tea Collection Inc. of San Francisco, Calif.
Importer: Tea Living Inc., d/b/a Tea Collection Inc. of San Francisco, Calif.
Month of recall: December 28, 2016
Country of origin: Thailand

Figure 7.24   Children's knit denim jackets

7.2.25   Children's nightgowns and two-piece pajama sets (Figure 7.25)

This recall involves children's 100% cotton woven, nightgowns and two-piece, long-sleeve top, and pant pajama sets. The nightgown has a peter pan collar with a red and white gingham pattern trim. The nightgown has six plastic buttons located on the back of the garment. The two-piece pajama set is traditionally styled with five plastic buttons on the center-front of the top with two pockets placed near the waist of the top. The pajama sets were sold in striped light blue, striped navy, striped red, striped pink, and lavender.[28]

Figure 7.25   Children's nightgowns and two-piece pajama sets

The children's nightgowns and two-piece pajama sets fail to meet the federal flammability standards for children's sleepwear posing a risk of burn injuries to children.

Stores responsible for the product: Children's boutique stores nationwide and online at www.dondolo.com.

Importer: Dondolo, of Carrollton, Texas
Month of recall: November 1, 2017
Country of origin: Colombia

7.2.26 Children's rompers (Figure 7.26)

This recall involves two styles of children's rompers sold in sizes 0-3 months and 18-24 months. They are Vermillion Painted OPP Floral Romper with

Figure 7.26   Children's rompers

style number 7F32500, and the Shocking Fuchsia Rose Romper with style number 7F32504. The Vermillion rompers are red with white floral print, and the Shocking Fuchsia are maroon with a pink floral print. The style number is printed on a tag sewn on the inside of the garment located in the waist area.[29]

The snaps near the collar can detach, posing a choking hazard to young children.

Stores responsible for the product: Nordstrom, Von Maur, and various boutique stores nationwide and online at teacollection.com

Importer: Tea Living Inc, San Francisco, Calif. (d/b/a Tea Collection)

Month of recall: February 13, 2018

Country of origin: Indonesia

7.3   Necessity of recall

Product recall is to limit liability for corporate negligence. A manufacturer that is supplying a hazardous product should recall it from the distributor immediately, if required to prevent cases of injury. If this measure is insufficient in preventing cases of injury, the manufacturer must without delay recall the product from those consumers that have purchased it. The recall should be carried out to the extent that is reasonable, considering the need to prevent cases of injury. In any case, recalls are always costly and add to the total quality cost[30] of a company because of replacing the recalled product or paying for damages caused in use. However, they are less costly than indirect cost that tarnishes image of a brand, reduced trust in the manufacturer, and faith in the consumer world.

References

1.   Gupta R K (2007), 'Product recalls: marketing failure and implications', Aravali Institute of Management, Jodhpur, Faculty column. Available from: http://www.indianmba.com/Faculty_Column/FC636/fc636.html [Accessed 1 March 2009].

2.   Emilia L. Sweeney (2008), 'Consumer protection: consumer product safety act: it's not a game', *Washington business magazine*, November/December issue, 2008.

3.   News from CPSC (2007), U.S. Consumer product safety commission 'Gap outlet recalls boys' jackets: drawstring at waist poses entrapment hazard', Recall release no 08-152. Available from: http://www.cpsc.gov/cpscpub/prerel/prhtml08/08152.html [Accessed 1 March 2009].

4.   News from CPSC (1996), U.S. Consumer product safety commission, 'Guidelines for drawstrings on children's upper outerwear', Available from: http://www.cpsc.gov/cpscpub/pubs/208.pdf [Accessed 1 March 2009].

5. News from CPSC (2007), U.S. Consumer product safety commission "TKS children's pants recalled by Sears; drawstrings at waist pose entrapment hazard", Recall release #08-116. Available from: http://www.cpsc.gov/cpscpub/prerel/prhtml08/08116.html [Accessed 1 March 2009].

6. News from CPSC (2007), U.S. Consumer product safety commission, 'Basic editions girls' clothing sets recalled by Kmart; drawstrings at waist pose entrapment hazard', Recall release #08-117. Available from: http://uihs.org/Product%20Recalls/08117.pdf [Accessed 2 March 2009].

7. News from CPSC (2007), U.S. Consumer product safety commission, 'Personal identity children's sweaters with drawstrings recalled by Sears due to strangulation hazard', Recall release #08-118. Available from: http://www.cpsc.gov/cpscpub/prerel/prhtml08/08118.htm [Accessed 2 March 2009].

8. News from CPSC (2007), U.S. Consumer product safety commission, 'Girls' hooded sweatshirts with drawstrings recalled by Liberty Apparel due to strangulation hazard', Recall release #08-146, Available from: http://www.cpsc.gov/cpscpub/prerel/prhtml08/08146.html [Accessed 2 March 2009].

9. News from CPSC (2008), U.S. Consumer product safety commission, 'Children's hooded sweatshirts recalled by urgent gear due to strangulation hazard; sold exclusively at Nordstrom stores', Recall release #08-217, Available from: http://www.cpsc.gov/cpscpub/prerel/prhtml08/08217.html [Accessed 2 March 2009].

10. News from CPSC (2008), U.S. Consumer product safety commission, 'Children's hooded sweatshirts recalled by Brents-Riordan Inc. due to strangulation hazard', Recall release #08-238, Available from: http://www.cpsc.gov/cpscpub/prerel/prhtml08/08238.html [Accessed 3 March 2009].

11. CBC news (2008), Consumer life, Children's products – recalls and advisories, 'Dora the explorer pajamas recalled'. Available from: http://www.cbc.ca/consumer/recalls/2008/05/dora_the_explorer_pajamas_reca_3.html [Accessed 3 March 2009].

12. News from CPSC (2008), U.S. Consumer product safety commission, 'Disney store recalls pirates of the caribbean sleeping bags due to violation of lead paint standard', Recall release #08-278. Available from: http://www.cpsc.gov/cpscpub/prerel/prhtml08/08278.html [Accessed 3 March 2009].

13. News from CPSC(2009) , U.S. Consumer product safety commission, 'Rashti & Rashti expands recall of infant garments due to choking hazard', Recall release #09-087, Available from: http://www.cpsc.gov/cpscpub/prerel/prhtml09/09087.html [Accessed 4 March 2009].

14. News from CPSC (2008), U.S. Consumer product safety commission, 'Children's board skirts recalled by Chelsea & Scott Ltd. due to violation of lead paint standard', Recall alert #08-599. Available from: http://www.cpsc.gov/cpscpub/prerel/prhtml08/08599.html [Accessed 4 March 2009].

15. News from CPSC (2008), U.S. Consumer product safety commission, "Circo children's Bobbie socks recalled due to choking hazard; sold exclusively at Target", Recall release #08-386. Available from: http://www.cpsc.gov/cpscpub/prerel/prhtml08/08386.html [Accessed 4 March 2009].

16. News from CPSC (2008), U.S. Consumer product safety commission, 'Hooded sweaters recalled by Empress Arts; children can strangle on drawstrings', Recall release #08-383. Available from: http://www.cpsc.gov/cpscpub/prerel/prhtml08/08383.html, [Accessed 5 March 2009].

17. News from CPSC (2008), U.S. Consumer product safety commission, 'CPSC and Carter's advise parents of Rashes Associated with Heat Transferred, or "Tag-less," Labels', Recall release #09-023. Available from: http://www.cpsc.gov/cpscpub/prerel/prhtml09/09023.html [Accessed 5 March 2009].

18. CBC news (2008), Consumer life, Children's products – recalls and advisories, 'Halloween costumes recalled due to fire danger'. Available from: http://www.cbc.ca/consumer/recalls/2008/10/2_halloween_costumes_recalled.html [Accessed 5 March 2009].

19. News from CPSC (2007), U.S. Consumer product safety commission, 'Nordstrom recalls children's flannel lounge pants due to burn hazard', Recall alert #07-533. Available from: http://www.cpsc.gov/cpscpub/prerel/prhtml07/07533.html [Accessed 13 March 2009].

20. News from CPSC (2008), U.S. Consumer product safety commission, 'JC Penney Recalls Arizona® Newborn and Infant Pants Due to Choking Hazard', Recall release #09-056. Available from: http://www.cpsc.gov/cpscpub/prerel/prhtml09/09056.html [Accessed 5 March 2009].

21. News from CPSC (2008), U.S. Consumer product safety commission, 'Doll Clothing Sets Recalled by Manhattan Group Due to Violation of Lead Paint Standard', Recall release #09-059. Available from: http://www.cpsc.gov/cpscpub/prerel/prhtml09/09059.htm [Accessed 5 March 2009].

22. News from CPSC (2008), U.S. Consumer product safety commission, 'Meijer Inc. Recalls Toddler Girl's Hat and Mitten Sets Due to Choking Hazard', Recall alert #09-711. Available from: http://www.cpsc.gov/cpscpub/prerel/prhtml09/09711.html [Accessed 5 March 2009].

23. News from CPSC (2008), U.S. Consumer product safety commission, 'Children's Hooded Jackets with Drawstrings Recalled by R&D International Due to Strangulation Hazard', Recall release #09-047, Available from: http://www.cpsc.gov/cpscpub/prerel/prhtml09/09047.html [Accessed 5 March 2009].

24. News from CPSC (2015), U.S. Consumer product safety commission, 'Precious Cargo infant one-piece garments' Recalled by SanMar Corp. of Issaquah, Wash, Recall release #15-208, Available from: https://www.cpsc.gov/Recalls/2015/precious-cargo-recalls-infant-one-piece-garments [Accessed 8 April 2018].

25. News from CPSC (2013), U.S. Consumer product safety commission, 'Girl's Three-Piece Clothing Sets' Recalled by Children's Apparel Network, of New York, N.Y., Recall release #13 - 177, Available from: https://www.cpsc.gov/ko/Recalls/2013/childrens-apparel-network-recalls-girls-clothing-sets-waist-belt-poses-risk-of [Accessed 8 April 2018].

26. News from CPSC (2016), U.S. Consumer product safety commission, 'Children's loungewear' Recalled by Eleanor Rose,of Natchez, Miss., N.Y., Recall release #16-135, Available from: https://www.cpsc.gov/Recalls/2016/eleanor-rose-recalls-childrens-loungewear [Accessed 8 April 2018].

27. News from CPSC (2016), U.S. Consumer product safety commission, 'Children's knit denim jackets' Recalled by Tea Living Inc., d/b/a Tea Collection Inc. of San Francisco, Calif., Recall release #17 - 060, Available from: https://www.cpsc.gov/Recalls/2016/tea-collection-recalls-childrens-denim-jackets [Accessed 8 April 2018].

28. News from CPSC (2017), U.S. Consumer product safety commission, 'Children's nightgowns and two-piece pajama sets' Recalled by Dondolo, of Carrollton, Texas., Recall release #18 - 021, Available from: https://www.cpsc.gov/Recalls/2017/dondolo-recalls-childrens-sleepwear [Accessed 8 April 2018].

29. News from CPSC (2018), U.S. Consumer product safety commission, 'Children's rompers' Recalled by Tea Living Inc, San Francisco, Calif. (d/b/a Tea Collection), Recall release #18 - 097, Available from: https https://www.cpsc.gov/Recalls/2018/tea-collection-recalls-childrens-rompers-due-to-choking-hazard [Accessed 8 April 2018].

30. Das Subrata (2008), 'Importance of Cost of quality in apparel sector', *Asian Text J*, 17, 57-58.

CHAPTER 8

Quality characterisation of various denim fabrics

Abstract

Denim fabrics remain one of the most favourite choices for the fashion world. This chapter first discusses different characters of denim fabrics by evaluating their mechanical and chemical properties. Raw denim fabric was subjected to various industrial washing treatments, such as whisker wash, river wash, caustic wash, potassium permanganate wash, snow wash, stone wash, bleach wash, acid wash, abrasion wash, and enzyme wash. Biopolishing of raw denim fabrics was also done with the enzyme. The chapter then discusses about the determination of weight, tensile strength, tearing strength, pH, colour fastness to washing, light, rubbing, and perspiration for different washed denim fabrics. Biopolished denim fabric was tested for weight loss due to abrasion, pilling, and colour change.

Keywords: denim, biopolishing, stone wash, whisker wash, snow wash, tensile strength, tearing strength

Chapter contains (Section headings)

8.1 Introduction
8.2 Washing treatments
    8.2.1 Whisker wash
    8.2.2 River wash
    8.2.3 Caustic wash
    8.2.4 Potassium permanganate wash
    8.2.5 Snow wash
    8.2.6 Stone wash
    8.2.7 Bleach wash
    8.2.8 Acid wash
    8.2.9 Abrasion wash
    8.2.10 Enzyme wash
    8.2.11 Biopolishing of denim fabric

8.1 Introduction

Industrial washing is one of the finishing methods applied on fabric or garment, which together with the use of new technologies and equipment enables to obtain the desired results. For finishing of denim fabrics, a range of treatment methods can be used. They all are aimed at new possible effects of fabric appearance; namely, mill wash or rinse wash, stone wash, moon wash, sand wash, bleach, over dyed look, damaged-look, scrubbed-look.[1] In recent years, there has been increasing interest in the use of environmentally friendly, nontoxic, fully biodegradable enzymes in the modern textile finishing process. Enzymatic treatment can replace a number of mechanical and chemical operations, which have been applied to improve the comfort and quality of fabrics.[2,3] Research and developmental studies in this area have been focused on applying enzymes on cellulose materials based on the cotton, linen, viscose, and their blends with synthetics fibres.[3,4] To improve fabric handle and other technological properties, the softeners are widely used in such finishing operations.[5]

Denim is very strong, stiff, and hard wearing woven fabric.[6] Denim is normally made out of cotton and twill weave fabric that uses coloured warp and white weft yarn are used for jeans, work clothes, and casual wear.[7] Denim is normally dyed with indigo, vat, and sulphur dyes. Among these, share of indigo is 67%. Indigo dyes are widely used for fashion dyeing; in denim, fibres dyed with indigo are not included in fibre-transfer examinations, remains as surface dyeing.[8] The most commonly denim washing methods are bleach wash, acid wash, enzyme wash, normal wash, stone wash, etc.[7,9] Among the washing methods, bleach method is widely used in industry; especially, for denim washing to achieve required colour shade by hypochlorite bleaching. Although chlorine is a harsh chemical, harmful to human health, causes corrosion to washing machine, and destructive to cotton. It may cause damage to cotton due to the decomposition of cellulose in the aqueous solution of hypochlorite bleach and losses its tensile strength; produces many decomposed product in bleach washing and pass into the effluent, where it causes environmental pollution.[9] However, bleaching treatments have been successful for achieving desirable colour shade and soft hand feels of cotton denim garments, and are still in use in denim washing industry.

Acid wash on denim jeans is becoming very popular due to its significant contrasts and attractive appearance in colour. It is a chemical wash process that stripped the top layer of colour and makes a white surface while the colour remained in the lower layer of the denim, giving it a faded look. It can be carried out on indigo and sulphur dyed base fabric.[10] In acid washing, basic chemicals are potassium permanganate and phosphoric acid or ortho-phosphoric acid which are used for soaking the pumice stone.[11] The soaked pumice stone is applied to modify the appearance of denim apparel. Denim wear has gained popularity all over the world. As a result, jeans wear is one of the most prominent apparel items in the world[12].

Among various washing methods, enzymatic method is widely used in the industry.[13,14] Enzyme treatment of cellulosic garments results in the degradation of cellulose,[15] yielding shorter chain cellulose polymers, and reduces its mechanical strength. It may be difficulty in controlling the activity of enzyme during the treatment process. However, it has been successful in improving fibre flexibility, desirable appearance, and soft handle of cotton denim garments.[16,17] Enzyme washing is an environmentally friendly treatment for denim. Organic enzymes are put on to the denim and eat away the cellulose in the cotton. Enzyme washing produces several effects, and can be noticeable in making seams, pockets, and hems and a salt and pepper colour effect may be produced. Studies on different stone and biostone washing of denim garments revealed that neutral cellulases produced a fabric with higher lightness and increase in enzyme added to back staining. The Denim treatment with 100% pumice stone alone is not so effective. However, a combination of 100% pumice stone along with cellulase showed a good wash down effect.

Table 8.1   Physical characteristics of Denim fabric

Fabric	Weave	Warp Count	Weft Count	EPI	PPI	Cover Factor	GSM
Raw Denim	3/1 Twill	50^s	40^s	87	74	18.87	198.7

Denim washing is an important part of garments manufacturing by which outlook, and comfort characteristics and fashion are changed or modified and vintage effect can be produced. Thats why now a days, the importance and demand of denim garments washing are increasing day by day. Different denim fabrics were washed using commercially available recipe for various applications in apparel industries. Characterisation of such denim fabrics was carried by evaluating mechanical and chemical properties.

Raw denim fabric was collected and analysed (Aadhi Colours, Tirupur, Tamil Nadu, India). Fabric particulars are shown in Table 8.1.

8.2   Washing treatments

Various washing treatments were carried out on raw denim fabrics in a commercial laundry. The flow of the processes is followed to get the desired wash down effect. Chemicals used in washing, such as acid, alkali, bleaching agent, wetting agent, potassium permanganate, silicon softener, and enzyme were of industrial grade. The size of pumice stones available for stone washing vary from 3 to 6 cm in diameter.

8.2.1   Whisker wash

Process flow of whisker wash is given below:

1. Raw denim
2. Crunch machine (10 mins–3 times)
3. Acid wash and bleach with balls (35–40 mins)
4. Enzyme (30 mins)
5. Silicon softner (30 mins)
6. Drying (40 mins) at 60°c
7. Finsihing
8. Inspection

8.2.2   River wash

Process flow of river wash is illustrated below:

1. Raw denim

2. Stone wash machine - rubber balls (30 mins)
3. Bleaching (30 mins)
4. Silicon softner (30 mins)
5. Drying (40 mins) at 60°c
6. Finsihing
7. Inspection

8.2.3 Caustic wash

Process flow of caustic wash is given below:

1. Raw denim
2. Caustic soda (30 mins) at 60°c
3. Neutralization
4. Drying (40 mins) at 60°c
5. Finsihing
6. Inspection

8.2.4 Potassium permanganate wash

Process flow of potassium permanganate wash is given below:

1. Raw denim
2. Emery sheet
3. Spray (PP)
4. Machine washing (30 mins)
5. Wetting agent (30 mins)
6. Silicon softner (30 mins)
7. Drying (40 mins) at 60°c
8. Finsihing
9. Inspection

8.2.5 Snow wash

Process flow of snow wash is given below:

1. Raw denim
2. Spray–pigment and PP (10mins)
3. Washing (30mins)
4. Silicon softner (30 mins)
5. Drying (40 mins) at 60°c
6. Finsihing
7. Inspection

8.2.6   Stone wash

Process flow of stone wash is given below:

1.   Raw Denim
2.   Stone Wash Machine–Pumice Stone (30 Mins)
3.   Silicon Softner (30 Mins)
4.   Drying (40 Mins) At 60°C
5.   Finsihing
6.   Inspection

8.2.7   Bleach wash

Process flow of bleach wash is given below:

1.   Raw denim
2.   Washing machine (30 mins)
3.   Bleach with steam (30 mins) at 120°c
4.   Neutralization
5.   Silicon softner (30 mins)
6.   Drying (40 mins) at 60°c
7.   Finsihing
8.   Inspection

8.2.8   Acid wash

Process flow of acid wash is given below:

1.   Raw denim
2.   Rotated in acid drum (10 mins)
3.   Neutralization (10 mins)
4.   Silicon softner (30 mins)
5.   Drying (40 mins) at 60°c
6.   Finsihing
7.   Inspection

8.2.9   Abrasion wash

Process flow of abrasion wash is given below:

1.   Raw denim
2.   Stone wash machine–pumice stone (30 mins)
3.   Manual rubbing or scraping
4.   Silicon softner (30 mins)
5.   Drying (40 mins) at 60°c

6.   Finsihing
7.   Inspection

8.2.10   Enzyme wash

Process flow of enzyme wash is given below:

1.   Raw denim
2.   Stone wash machine - rubber balls (30 mins)
3.   Enzyme wash (30 mins)
4.   Water wash
5.   Silicon softner (30 mins)
6.   Drying (40 mins) at 60°c
7.   Finsihing
8.   Inspection

8.2.11   Biopolishing of denim fabric

Raw denim fabric was biopolished using the following recipe and processing condition:

Cellulase Enzyme   : 6%
M:L   : 1:30
Temperature   : 50°C
Time   : 60 minutes
pH   : 4 to 5 (Acetic Acid)

Processing sequence is given below:

1.   Biopolishing
2.   Washing
3.   Drying

8.3   Evaluation

Prior to test, various denim specimens were conditioned to moisture equilibrium in the standard atmosphere of 65±2% RH and 21±2°C temperature.

8.3.1   Fabric weight (Gram per square meter)

Denim fabric samples were evaluated for weight using the standard test method ISO 3801-1977. GSM cutter was used to cut the fabric, and weight was measured in calibrated electronic balance.

8.3.2   Fabric tensile strength

Tensile strength measurement was carried out by following ISO 13934-2:2014 test method. Five specimens were cut in warp and weft direction and mounted between two clamps of the Instron Tensile Testing machine in such a manner that the same set of yarns was gripped by both the clamps and continual increasing load was applied longitudinally to the specimen by moving one of the clamps until the specimen ruptures. Values of breaking load and elongation of the test specimen were noted from the machine.

8.3.3   Fabric tearing strength

Tearing strength measurement was carried out by following ISO 13937-1:2000 test method. Elmendorf Tearing Strength Tester was used and five warp way, and five weft way samples were tested. Average values of warp and weft were reported.

8.3.4   Determination of pH

All the washed denim samples were evaluated for pH by following AATCC 81:2006 test method. The specimens were boiled in distilled water. The water-extract was cooled to room temperature, and the pH was determined.

8.3.5   Colour fastness to washing

Washed denim samples were tested for colour fastness to washing by following ISO 105 C06-A2S test method. Launder-o-meter washing test instrument was used for the study. The change in colour of the specimen and the staining of the adjacent fabric were assessed by comparison with the grey scales.

8.3.6   Colour fastness to rubbing

Dry and wet rubbing of the specimens were carried out using ISO 105 X12 test method, and Crock meter was used for the testing. Samples were evaluated for colour staining on the crock square cloth using grey scale for staining.

8.3.7   Colour fastness to light

Colour fastness to light was determined using the ISO 105 B02 test method. A specimen of the denim sample to be tested was exposed to artificial light in Xenon Light Fastness Tester under controlled conditions, together with a set of reference materials. The colour fastness was assessed by comparing the change in colour of the test specimen with that of the reference blue wool standard.

8.3.8   Colour fastness to perspiration

Specimens were tested by following the test method ISO 105 E04 in both alkaline and acidic condition. Procedure of the preparation of alkaline and acidic solution is given below:

Alkaline solution

0.5 g of L-histidine monohydrochloride monohydrate ($C_6H_9O_2N_3 \cdot HC_1 \cdot H_2O$); 5 g of sodium chloride (NaCl);

The solution was brought to pH 5.5 ± 0.2 with 0.1 sodium hydroxide solution (pH adjustment).

Acid solution

0.5 g of L-histidine monohydrochloride monohydrate ($C_6H_9O_2N_3 \cdot HC_1 \cdot H_2O$); 5 g of sodium chloride (NaCl); 2,2 g of sodium dihydrogen orthophosphate dihydrate ($NaH_2PO_4 \cdot 2H_2O$).

The solution was brought to pH 5.5 (±0.2) with 0.1 sodium hydroxide solution (pH adjustment).The change in colour of each specimen and the staining of the adjacent fabrics were assessed by comparison with the grey scales.

8.4   Results and discussions

Ten different denim washed fabrics were analysed for fabric weight, tensile strength, tearing strength, pH, colour fastness to washing, colour fastness to rubbing, colour fastness to light, and colour fastness to perspiration. Raw denim fabric was also biopolished using cellulase enzyme in acidic medium and evaluated for weight loss, abrasion resistance, and pilling.

8.4.1   Fabric weight

It has been observed from the Table 8.2 that the fabric GSM varies from 189.2 to 197.1. Weight loss varies from 0.8% to 9.50%. Weight loss of fabric in abrasion wash is more because of higher abrasion of fibres and subsequent removal from the surface of the fabric using emery sheets. Weight loss of fabric in whisker wash is less because of the use of low concentration of chemicals, such as enzyme, acid, and bleaching agents.

8.4.2   Tensile strength

It has been observed from the Table 8.3 that the tensile strength varies from 47.1 to 56.38 lbs and 39.9 to 47.08 lbs in warp and weft direction, respectively, for different washing treatments. Strength loss varies from 2.69% to 18.70% and 2.29% to 15.25% in warp and weft direction, respectively.

Table 8.2   Effect of denim washing on fabric weight loss

Sl.No.	Wash Types	Fabric Weight (GSM)	% Weight Loss
1	Raw denim	198.7	–
2	Whisker wash	195.4	1.66
3	River wash	194.7	2.01
4	Caustic wash	196.5	1.11
5	PP wash	193.8	2.45
6	Snow wash	196.2	1.26
7	Stone wash	197.1	0.8
8	Bleach wash	192.3	3.22
9	Acid wash	193.0	2.87
10	Abrasion wash	189.2	9.50
11	Enzyme wash	194.2	2.26

Table 8.3   Effect of denim washing on tensile strength

Sl.No.	Wash Types	Tensile Strength (lbs)		% Strength Loss	
		Warp	Weft	Warp	Weft
1	Raw denim	57.94	47.08	–	–
2	Whisker wash	56.38	46.0	2.69	2.29
3	River wash	51.62	43.12	10.91	8.41
4	Caustic wash	51.64	42.3	10.87	10.15
5	PP wash	52.58	43.18	9.25	8.28
6	Snow wash	51.74	42.12	10.70	10.53
7	Stone wash	53.84	44.16	7.07	6.20
8	Bleach wash	54.58	45.2	5.79	3.99
9	Acid wash	48.9	41.22	15.60	12.44
10	Abrasion wash	47.1	39.9	18.70	15.25
11	Enzyme wash	50.14	42.06	13.46	10.66

Maximum strength loss has been obtained for abrasion wash in both warp and weft direction. Minimum strength loss has been obtained for whisker wash in both warp and weft direction. Strength loss for abrasion wash is high because of higher abrasion of the surface of the fabric using emery sheets. The strength loss for whisker wash is low because of the use of low concentration of chemicals, such as enzyme, acid, and bleaching agents.

It was observed from Figure 8.1 and Figure 8.2 that with increase in weight loss of different washing treatments of denim fabrics, there is increase in strength loss in both warp and weft direction. This may be due to the fact that weight loss leads to the removal of fibres from the lattice structure of fabrics which results in increase in strength loss.

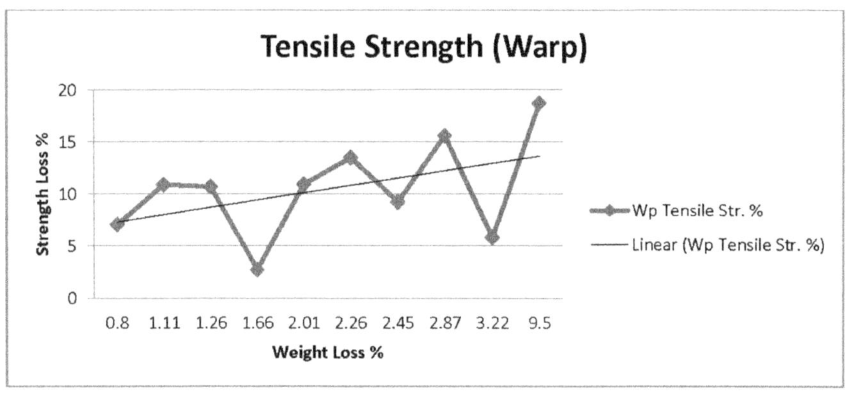

Figure 8.1   Effect of weight loss on tensile strength (Warp) loss

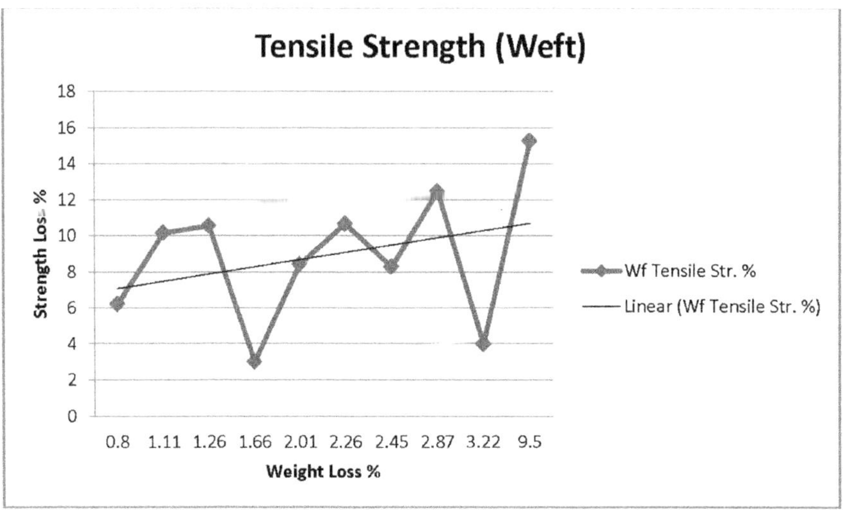

Figure 8.2   Effect of weight loss on tensile strength (weft) loss

8.4.3 Tearing strength

It has been observed from Table 8.4 that the tearing strength varies from 2995.2 to 3635.2 gf and 2419.2 to 2739.2 gf in warp and weft direction, respectively, for different washing treatments. The strength loss varies from 2.73% to 19.86% and 1.86% to 12.46% in warp and weft direction, respectively. Maximum tearing strength loss has been obtained for abrasion wash in warp direction. Maximum strength loss has been obtained for acid wash in weft direction. Minimum strength loss has been obtained for whisker wash in both warp and weft direction.

Tearing strength loss for abrasion wash is high because of more abrasion on the surface of the fabric and subsequent weakening of fibre assembly using emery sheets. Tearing strength loss for whisker wash is low because of the use of low concentration of chemicals, such as enzyme, acid, and bleaching agents.

It is observed from Figure 8.3 and Figure 8.4 that with increase in weight loss % of different washing treatments of denim fabrics, there is increase in tearing strength loss in warp direction due to weakening of fibrous assembly. However, no change is noticed in the weft direction.

Table 8.4 Effect of denim washing on tearing strength

Sl.No.	Wash Types	Tearing Strength (Gram Force)		% Strength Loss	
		Warp	Weft	Warp	Weft
1	Raw denim	3737.6	2739.2	–	–
2	Whisker wash	3635.2	2688	2.73	1.86
3	River wash	3225.6	2560	13.69	6.54
4	Caustic wash	3584	2662.4	4.10	2.80
5	PP wash	3443.2	2585.6	7.87	5.60
6	Snow wash	3264	2419.2	12.67	11.68
7	Stone wash	3520	2649.6	5.82	3.27
8	Bleach wash	3392	2739.2	9.25	2.33
9	Acid wash	3136	2624	16.09	12.64
10	Abrasion wash	2995.2	2688	19.86	1.87
11	Enzyme wash	3328	2649.6	10.95	3.27

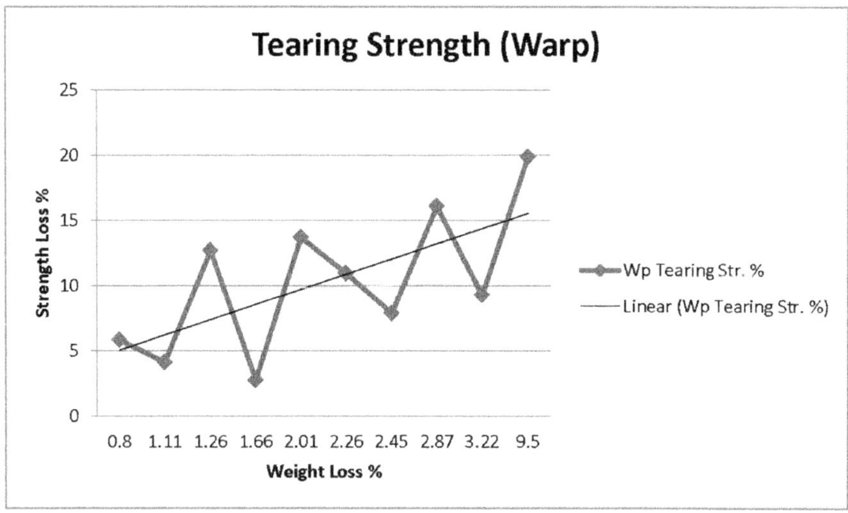

Figure 8.3 Effect of weight loss on tearing strength (warp) loss

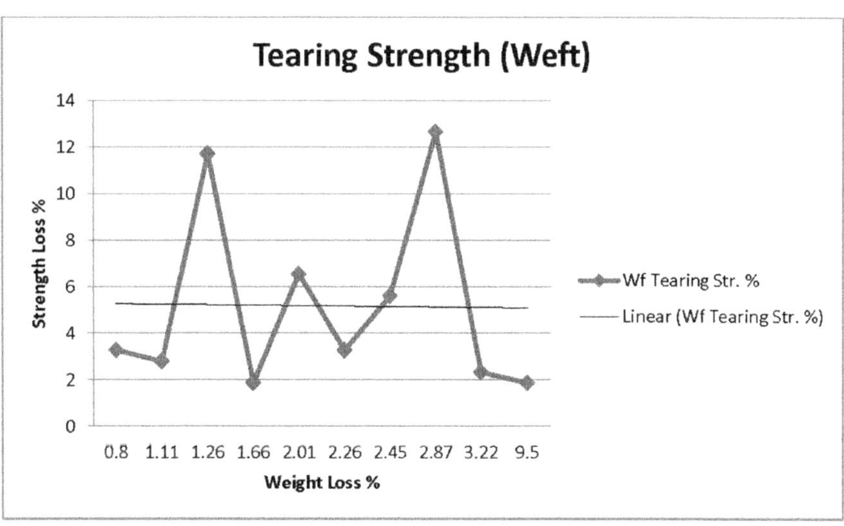

Figure 8.4 Effect of weight loss on tearing strength (weft) loss

8.4.4 pH value

It has been found from Table 8.5 that the pH varies from 8.10 to 8.55 for different denim fabrics, and there is no significance difference between the pH values of different washing techniques. Highest value of pH was observed in whisker wash (8.55), and the lowest value of pH was found in bleach wash (8.10).

Table 8.5.   Effect of denim washing on pH

Sl. No.	Wash Types	pH Value
1	Raw denim	8.64
2	Whisker wash	8.55
3	River wash	8.42
4	Caustic wash	8.27
5	PP wash	8.24
6	Snow wash	8.31
7	Stone wash	8.33
8	Bleach wash	8.10
9	Acid wash	8.28
10	Abrasion wash	8.35
11	Enzyme wash	8.17

8.4.5   Colour fastness to washing

Colour fastness to washing for all the categories of denim fabrics is shown in Table 8.6. It is corroborated from the observed values that, change in colour and colour staining on multifibres, do not vary with different types of washes on denim fabrics.

Colour staining rating on Nylon is lower i.e. 3–4, in comparison to colour staining on other multifibres. Change in colour in whisker wash is lower when compared with other washes on denim fabrics.

8.4.6   Colour fastness to rubbing

It has been found from the Table 8.7 that the rubbing fastness rating in wet condition is 2–3 for river wash, caustic wash, and enzyme wash and in dry condition rating is 3–4 for river wash, caustic wash, PP wash, abrasion wash, and enzyme wash. It may be due to the presence of residual unfixed dyes on the surface of the washed denim fabrics in comparison to other washes.

8.4.7   Colour fastness to light

It has been observed from the Table 8.8 that the colour fastness to light does not vary with different types of washes on denim fabrics. All the washes in denim fabrics in colour fastness to light exhibit colour fastness rating of 3.

Table 8.6 Effect of denim washing on colour fastness to washing

Sl. No.	Wash Types	Change In Colour	Colour Staining On Acetate	Colour Staining On Cotton	Colour Staining On Nylon	Colour Staining On Polyester	Colour Staining On Acrylic	Colour Staining On Wool
1	Raw denim	4	4	4	4	4	4	4
2	Whisker wash	3–4	4	4	3–4	4	4	4
3	River wash	4	4	4	3–4	4	4	4
4	Caustic wash	4	4	4	3–4	4	4	4
5	PP wash	4	4	4	3–4	4	4	4
6	Snow wash	4	4	4	3–4	4	4	4
7	Stone wash	4	4	4	3–4	4	4	4
8	Bleach wash	4	4	4	3–4	4	4	4
9	Acid wash	4	4	4	3–4	4	4	4
10	Abrasion wash	4	4	4	3–4	4	4	4
11	Enzyme wash	4	4	4	4	4	4	4

Table 8.7   Effect of denim washing on colour fastness to rubbing

Sl. No.	Wash type	Colour Fastness Dry	Colour Fastness To Rubbing Wet
1	Raw denim	4	3
2	Whisker wash	4	3
3	River wash	3–4	2–3
4	Caustic wash	3–4	2–3
5	PP wash	3–4	3
6	Snow wash	4	3
7	Stone wash	4	3–4
8	Bleach wash	4	3
9	Acid wash	4	3
10	Abrasion wash	3–4	4
11	Enzyme wash	3–4	2–3

Table 8.8   Effect of denim washing on colour fastness to light

Sl.No.	Wash Types	Colour Fastness to Light
1	Raw denim	3
2	Whisker wash	3
3	River wash	3
4	Caustic wash	3
5	PP wash	3
6	Snow wash	3
7	Stone wash	3
8	Bleach wash	3
9	Acid wash	3
10	Abrasion wash	3
11	Enzyme wash	3

8.4.8   Colour fastness to perspiration

Colour fastness to perspiration is shown in the Tables 8.9a and 8.9b. It has been found from the tables that there is no variation in colour change and colour staining on multifibres except nylon. Colour staining on Nylon is 4–5 and 4 in acidic and alkali medium, respectively.

Table 8.9a Effect of denim washing on colour fastness to perspiration (acid)

Sl.No.	Wash Types	Change In Colour	Colour Staining On Acetate	Colour Staining On Cotton	Colour Staining On Nylon	Colour Staining On Polyester	Colour Staining On Acrylic	Colour Staining On Wool
1	Raw denim	4	4–5	4	4	4–5	4–5	4–5
2	Whisker wash	4	4–5	4	4	4–5	4–5	4–5
3	River wash	4	4–5	4	4	4–5	4–5	4–5
4	Caustic wash	4	4–5	4–5	4–5	4–5	4–5	4–5
5	PP wash	4	4–5	4	4	4	4–5	4–5
6	Snow wash	4	4–5	4	4	4	4–5	4–5
7	Stone wash	4	4–5	4	4	4–5	4–5	4–5
8	Bleach wash	4	4–5	4	4	4–5	4–5	4–5
9	Acid wash	4	4–5	4	4	4–5	4–5	4–5
10	Abrasion wash	4	4–5	4	4	4–5	4–5	4–5
11	Enzyme wash	4	4–5	4	4	4–5	4–5	4–5

Table 8.9b   Effect of denim washing on colour fastness to perspiration (alkali)

Sl.No.	Wash Types	Change In Colour	Colour Staining On Acetate	Colour Staining On Cotton	Colour Staining On Nylon	Colour Staining On Polyester	Colour Staining On Acrylic	Colour Staining On Wool
1	Raw denim	4	4–5	4	4	4–5	4–5	4–5
2	Whisker wash	4	4–5	4	4	4–5	4–5	4–5
3	River wash	4	4–5	4	4	4–5	4–5	4–5
4	Caustic wash	4	4–5	4–5	4	4–5	4–5	4–5
5	PP wash	4	4–5	4	4	4	4–5	4–5
6	Snow wash	4	4–5	4	4	4–5	4–5	4–5
7	Stone wash	4	4–5	4	4	4–5	4–5	4–5
8	Bleach wash	4	4–5	4	4	4–5	4–5	4–5
9	Acid wash	4	4–5	4	4	4–5	4–5	4–5
10	Abrasion wash	4	4–5	4	4	4–5	4–5	4–5
11	Enzyme wash	4	4–5	4	4	4–5	4–5	4–5

8.5   Biopolished denim fabric

Comparative assessment between ordinary raw denim fabric and biopolished denim fabric are made. It has been found from Table 8.10 that after biopolishing, there is reduction in abrasion weight loss, improvement in pilling, and colour change. It is due to the controlled hydrolysis of cellulosic fibres in denim fabric with the application of cellulase enzyme resulting in improvement in pilling rating, colour change, and reduction in abrasion weight loss.

8.6   Conclusions

- Different washed denim fabrics have unique characteristics, and they can be used in different apparel products.
- Whisker wash was found to be better in mechanical and colour fastness properties because of the use of low concentration of chemicals, such as enzyme, acid, and bleaching agents.
- River wash, caustic wash, PP wash, snow wash, stone wash, bleach wash, acid wash, and enzyme wash show moderate result in mechanical and colour fastness properties. Moderate colour fastness properties may be due to the presence of residual unfixed dyes on the surface of washed denim fabric in comparison to other washes.
- Abrasion wash has poor mechanical properties in comparison to other denim washed fabrics because of higher surface abrasion of the fabric using emery sheets.
- Improvement in pilling, colour change, and weight loss due to abrasion were noticed in biopolished denim fabric.

Table 8.10:   Effect of biopolishing on denim fabric

Fabric	No. of Rubs	Abrasion Weight Loss (%)	Pilling	Colour Change
Raw denim	150	4.09%	3–4	3
Biopolished denim	150	2.8%	4	3–4

References

1. Anon (2016), 'Enzymes'. Available from: http://www.cht-group.com/(accessed December 15, 2016).

2. Anon (2016), 'Enzymes for Textiles'. Available from: http://www.mapsenzymes.com/ Enzymes_Textile.asp (accessed December 17, 2016).

3. Buschle-Diller, G., Walsh, W. K., Radhakrishnaiah, P. (1997), 'Effect of Enzymatic Treatment on Dyeing and Finishing of Cellulosic Fibers: A Study of the Basic Mechanisms and Optimization of the Process', Project: C96-Al National Textile Center Annual Report: November, 31–36.

4. Onar, N., Saruşik, M. (2005), 'Use of Enzymes and Chitosan Biopolymer in Wool Dyeing', *Fibres & Textiles in Eastern Europe*, 49 (1), 54–59.

5. Buscle-Diller, G., Dong Yang, X. (2001), 'Enzymatic Bleaching of Cotton Fabric with Glucose Oxidase', *Textile Res J.*, 71 (5), 388–394.

6. Razzaque, M. A. (2004), Garment & Textile Merchandising, 1st edition. Popular Publications, Dhaka, Bangladesh, 223-226.

7. Kashem, M. A. (2008), Apparel Merchandising, 1st edition. Lucky-One Traders, Dhaka, Bangladesh, 69-71.

8. Grieve, M., Biermann, T., & Schaub, K (2006), 'The Use of Indigo Derivatives to Dye Denim Material', *Science & Justice*, 46, 15–24.

9. Islam, M. T. (2010), Garments Washing & Dyeing, Ananto Publications, Dhaka, Bangladesh, 220–222.

10. Arjun, D., Hiranmayee, J., Farheen, M.N.(2013), 'Technology of Industrial denim washing: Review' *International journal of industrial engineering &Technology*, 3 (4), 25–34.

11. Hams Group (2009), Personal communication, Hams Washing & Dyeing Limited, Tejgaon I/A, Dhaka, Bangladesh..

12. Khan, M.E. (2011), Technology of denim manufacturing, 1st Edn. Books fair publication, Dhaka, Bangladesh, 1–2.

13. Buchert, J. and Heikinheimo, L. (1998), 'New cellulase processes for the textile industry', *Carbohydr. Eur*, 22, 2–4.

14. Duran, N. and Marcela, D.(2000), 'Enzyme applications in the textile industry', *Review Progress in Colouration*, 30(1), 41–44.

15. Morries, C. E. and Harper, R. J. (1994), 'Comprehensive view on garment dyeing and finishing', *American Dyestuff Reporter*, 83, 132–136.

16. Heikinheimo, L., and Buchert, J., Miettinen-oinonen, A., Suominen, P. (2000), 'Treating Denim Fabrics with Trichoderma Reesei Cellulases', *Textile Res. J*, 70, 969–973.

17. Harrison PW (1988), 'Garment dyeing (Ready to wear Fashion from dye house)', *Textile Progress*, The Textile Institute, UK. 19 (2).

CHAPTER 9

Characterisation of leather and leather garments

Abstract

Application of leather in apparel sector is gaining importance due to fashion trend and subsequent increase in demand from the industry. Different kinds of leather hide are used in leather garment. In this chapter, characteristic of such leather hides are discussed. The use of special finishes in leather is very common to make it attractive and accepted for different end uses. Essential features of such finishes have been the focus point in the apparel sector. Thus, the quality characterization of different finished leather has been highlighted. Finally, performance requirements of different leather and suede garment have been discussed. Some of the specific tests for leather items are also been indicated.

Keywords: hide, shearling, nubuck finish, cracking, scuff resistance, stitch tear strength

Chapter contains (Section headings)

9.1 Introduction
9.2 Characteristics of different kinds of leather hides
    9.2.1 Cowhide
    9.2.2 Lambskin
    9.2.3 Sheepskin
    9.2.4 Pigskin
    9.2.5 Shearling
    9.2.6 Sheep or lamb
    9.2.7 Goatskin
9.3 Types of finishes in leather garments
    9.3.1 Aniline finish
    9.3.2 Semi-aniline finish
    9.3.3 Antique finish

9.1 Introduction

Leather was the first clothing fabric over the past few decades. Leather has developed a racing reputation from bickers collars to lingerie. Leather is emerging as a hip look on the streets and in the office covering men or women, young, or old from top to bottom. Due to a blend of nature and modern technology, the new look of leather is soft and supple. Leather has become a very specialized high fashion fabric that requires talented specialists to turn into a quality garment. Leather is tailored into hipster sheep skin pants, clinched waist jackets, goatskin shirts, double breasted suede shearling with shawl collar in a petite fit, pig suede soft silk nap sportswear, etc. Clothing leather is thin, versatile and almost silky and comes in a variety of eye catching colours viz. red, camel, gold metallic, or olive. Even blue or green are no strangers in the rainbow shades. Animal prints, which are so popular in clothes, come to life as leather outfits. Relative to virtually all manmade textiles, leather is very strong and has a high resistance to tears and punctures. The comfort provided by leather garments is due to leather's ability to combine breathing and insulating properties. Leather is hot in summers and cold in winters, but in reality, it adjusts constantly to its environment. This is because, leather is a natural product and it breathes freely maintaining a comfort level in all seasons. Even in warm climates, leather is wearable and bearable. Leathers are constantly improved, as new finishes and colours are created. The richness and variety of leathers available to everyone at reasonable prices would be the envy of ancient potentates. Leathers and suedes can be dyed, glazed, buffed, polished, foiled, embossed, printed, patented, beaded, sparkled, perforated, stenciled silk screened, woven fringed or embroidered, to create a variety of looks. Leather is tanned and finished so that the grain side has a smooth rich surface.

With leather garments so popular in the high street, it is critical to ensure that the performance of these products meets customer expectations. Despite many leather garments being price point items, customers still have a perception of quality because they are made of leather. To ensure that leather garments perform as expected there exist a range of tests, designed to test the garments' suitability for wear, including finish adhesion, tear strength, colour fastness, pH, rub fastness, etc. One of the most common consumer complaints is damage due to drycleaning. It is no longer enough to label garments as "specialist dry clean only". There are several different cleaning methods all of which can cause different reactions with the leather.

9.2 Characteristics of different kinds of leather hides[1]

Most common types of leathers which are used for making leather garments is given below:

9.2.1   Cowhide

Cowhide is the most common leather used in the making of garments. It covers a wide spectrum of textures and quality. It is quite durable, easy to care for, and resistant to water and dirt. It takes the shape of the wearer making it more comfortable with every day use. This affordable functional leather offers fashion, value and endless colours, and style. Cow napa leathers are suitable for making leather garments. Cow napa leathers are preferred in garments making because of its large area with increased cutting value and smoothness of grain. It has good drape, fluffy feel, lightweight, cold-crack resistance and good stitch tear, and tongue resistance.

9.2.2   Lambskin

Lambskin is very soft, luxurious leather. Its natural lightweight layers give it distinctive velvet touch, which suits fitting jackets, pants, skirts as well as coats. With a little extra care, lambskin is very wearable and the ultimate luxury.

9.2.3   Sheepskin

Sheepskin refers to the hide of a sheep used with the wool still attached. Usually, the wool side faces into the garment or accessory, but it can also be reversible. The wool can be ironed which means straightened to yield a smooth fur like appearance or it can be left naturally curly. Irrespective of the way the wool is styled, this is the warmest leather available.

9.2.4   Pigskin

Pigskin is the most popular and versatile, easily transformed into fashion's most current look. When tanned outside, it provides smooth napa finish, often used for jackets and accessories. Tanning on the inside results in a silky suede finish. The natural lightweight structure of the skin produces delicate patterns, textures, and silky soft naps, perfect for sportswear, shirts, and blazers.

9.2.5   Shearling

Shearling refers to hides from lambs, which are much lighter in weight than sheepskin and much softer. Although they may be lighter, they are just as warm as the heavier sheepskins. They are elegant and attractive to a fur coat. Shearling hides has gone through a limited shearing process to obtain a uniform depth of the wool fibres for a uniform look and feel. However, shearling is not shorn wool; the term refers to the pelt of a yearling sheep that has been shorn only once by the process described above. Shearling garments are made from the pelts by tanning them with the wool of uniform depth still on them. A typical shearling pelt has leather, on one side and shorn fibres on the other side. The fibres of shearling tend to wick away moisture or retain moisture, depending on humidity to give comfort to the wearer. The wool and hide can be dyed in any colour, although black is the most common. Shearling coats have been worn since ancient times throughout Asia and Europe. They remain popular in many countries, although they are made in modern styles.

9.2.6 Sheep or lamb

It is relatively unblemished skin. Its particular fibre structures and natural characteristics blend itself perfectly in today's fashion leather garments. Sheepskins are the unique material for making garments because of the real feather-touch feel, good softness, and smooth grain when compared with the garments made from the goatskins and cowhides. Sheep napa is the ideal raw material for making leather garments, and it has the quality to drape well, fluffy feel, light weight, fashion feel, wear resistance, good strength, uniform shade, glossy appearance and thickness, wash resistance, fastness to wet and dry rubbing, fastness to perspiration, shower proof resistance, resistance to heat (ironing) cold and light.

9.2.7 Goatskin

Goatskin has characteristic pattern and short and compact fibrous structure, with a unique look to the skin. It is used for a variety of products ranging from gloves to garments.Goatskins produce finer suede when compared with cow hides. The suede has sheen nap which is uniform, tightly packed, and resilient.

9.3 Types of finishes in leather garments[1]

9.3.1 Aniline finish

Aniline leathers, which are tumbled in vats, completely absorb the dye. There is no other colouring agents or process. Thus, the finished leather tends to look and feel more natural. The unique markings and characters of each skin are apparent. Usually, the best quality of sides are reserved for this process as aniline leathers are valued highest of consumers.

9.3.2 Semi-aniline finish

Semi-aniline leathers are the combination of both pigment and aniline dyed. A very light pigment is added to even out the colour and increase the durability. Most garments are made with semi-aniline leathers.

9.3.3 Antique finish

The light application of one colour over another colour (usually, a darker colour over the lighter colour) is to create highlights.

9.3.4 Nubuck finish

The leather is finished with velvet-like surface on the grain. Because the fibres in the grain layer are compact and short, the nap is fine and smooth. It is often mistaken for suede, but suede is the flesh side of the piece of leather while nubuck is an effect that is done to the grain side, making it considerably smooth and stronger.

9.3.5 Oil-pull up finish

Oil pull up is done with the application of oils in full grain leathers. The oil can migrate when pressure is applied on the surface and come back when the pressure is released. Thus, the surface will show two-tone effect when pressed or pulled.

9.3.6 Corrected grain finish

The crust is buffed to remove the top grain pattern and treated with a filling type of resin binder, which makes the grain layer tight. Afterwards, several coating of finish mixes with high amount of covering material is applied. An artificial pattern similar to animal grain is applied using suitable pressure and heat in the hydraulic press. The finish can be modified to obtain glossy surface, waxy surface, etc. and the surface is relatively uniform. Special effects, such as brush off can be obtained using suitable binders and colouring mixes

9.3.7 Other special finishes

Other special finishes, such as stone wash, sand wash, destroyer, marble, denim, distressed, metallic, etc. are also done in garment leathers for making garments with high fashion and style.

9.4 Commercial Colour Match

Good looking leathers in subtle pastel colours, moulded by meticulous sewing methods add up to the stuff of delicate dreams in garment designing. In garments designing, the sensuous silhouettes, fine detailing, unconventional style, and invigorating colours add spark to the romantic feel. Flaunt apparels in a riot of colours are inspired by the exotica to create stunning looks. The right kind of colours enhances the beauty and lends dignity. Smart and sober, subdued and extravagant, and shapes and styles with a play of colours have a bohemian touch. Colour match should hold well under different light sources *viz.* daylight, domestic light, and even at the fluorescent light at point of sale. The Garment Industry must be able to provide consistently coloured garments, which will satisfy the end user and assess what is a "commercial "colour match given the differing production condition. Visual techniques of colour matching and colour assessment result in off shade garments being manufactured. Therefore, the colour assessment must be done with modern technology of using micro-processing system.

9.5   Selection of leathers

The selection of leathers is done according to buyers or customers' choice. The selection of leathers for display purposes is done according to the design, style, colour, and fashion trend.

9.6   Selection of linings

Selecting lining fabric for leather garment is very important. The lining is the finishing touch, covering the inner construction, and protecting the garments from unnecessary wear. It also absorbs the most of the wearing strain and prevents the garment from stretching out of shape; particularly, where closely fitted. Linings should be as durable as the garment leather. Lining should be of high quality that will withstand repeated wearing and drycleaning. Rayon, rayon/acetate, and polyester are excellent choice for linings. The linings should be lighter in weight and softer than the garment leather, with a smooth surface that slides over the garment without rustling. Lining should be chosen that falls into soft folds when draped over the hand. The textured linings must be avoided that cling to other garments. Although a jacket or coat lining is on the inside of the finished garment, the edges of the sleeve lining often show or the lining often visible on moving. Therefore, the colour is an important consideration. The lining colour must be dark enough to cover seam and hem edges and inner construction, but not so dark that it shows through to the right side of the garment. Generally, lining materials, such as nylon cotton, polyester cotton, silk, etc are used in leather garments.

9.7   Selection of accessories

Selection of accessories, such as buttons, belt buckles, hooks and eyes, zips, etc are done according to buyers or customers choice. The selection of accessories needed for display purposes is done according to the design and style of the garments.

9.8   Quality characterization of finished leathers[2]

Leather being a utility and fashion item, the customer expects the quality of the leather in terms of durability, aesthetic appearance, and fashion appeal. The quality of leathers is measured in terms of properties by visual examination. This includes smoothness of grain, colour, uniformity of dyeing, crackiness of grain, softness, fullness, and any other defects. The quality of leather is also measured in terms of physical properties, such as tensile strength, tear strength, grain crack resistance, water vapour permeability, rub fastness, finish adhesion strength, light fastness, abrasion resistance, etc. Because finished leathers are the raw materials for fabricating leather products, it is necessary to test the quality of the leathers before fabricating. Some of the important tests for selecting good quality leathers are:

9.8.1   Feel

The feel of leather is tested by feeling up in the palm at different places.

9.8.2   Adhesion to finish

Scotch tape of 5" is stuck to about 4" on the grain side of the leather and then pressed well. The loose end of the tape is ripped off in one quick motion. The tape is checked to see if any finish is sticking to the tape from the stuck surface of the leather. Poor adhesion of finish can give rise to flaking and peeling of finish in use.

9.8.3   Cracking

The leathers are double folded at least in four places to see whether there is a tendency of pigment or grain cracking.

9.8.4   Dry and wet rub

The grain side is rubbed vigorously with a piece of white fabric, and the cloth is examined for any transfer of finish after rubbing. The cloth is wet and then rubbed on the grain side at some place to find out any transfer of finish on the wet cloth. This involves the assessment of the change in shade of the leather after testing. The degree of change of colour is assessed using grey scales.

Colour fastness to rubbing – dry 150 rubs – min. 3 grade for upper - wet 50 rubs – min. 3 grade for upper leather - dry 50 rubs – min. 3 grade for suede leather - wet 20 rubs – min. 3 grade for suede leather - dry 50 rubs – min. 3 grade for napa leather - wet 20 rubs – min. 3 grades for napa leather.

9.8.5 Scuff resistance

The resistance of leather to damage under impact is measured using scuff resistance test.

9.8.6 Strength

A small cut is made on the butt region, and the leather is torn with fingers by pulling strongly on either side. The effort is compared with that required reference of strength. (Min 35-40 kg/cm).

9.8.7 Fading

In case of white or light coloured leather, the leather piece is kept in the sun for 3 hours and compared it with the original piece for fading (yellow or change of colour). The performance of leather surface to resist fading to light is compared with a standard wool scale to give a performance rating. Pigment finished systems will usually perform better than aniline dye type finishes.

9.8.8 Water absorption

A small piece of the leather is dipped in water for 5 minutes and **Bally Permeometer** measures the amount of water absorbed by it. The grain side of the piece is rubbed with a piece of fabric to see physical determinants. Water absorption test is carried out for both upper and garment leathers using **Penetrometer** and the standards prescribed for water absorption test are strictly followed.

Bally Permeometer and Penetrometer for upper and garment leathers to water:

Penetration time - minimum 60 minutes for upper.

Water absorption percentage – maximum 80% for 60 minutes for upper (the above test is not applicable to garment leathers).

9.8.9 Water repellent

This test is mainly applicable to clothing leather and measures the shower proof. It is carried out by spraying water on a piece of leather under controlled conditions and comparing the effect with a standard.

9.8.10 Chemical resistance

Some water is dropped on the grain side of a piece of leather and dried. Later natural rubber adhesive, synthetic neoprene rubber adhesive, etc are applied and allowed to dry. The piece is then rubbed with a crepe rubber to see any damage to the finish.

9.8.11 Tensile strength

It is one of the important properties to evaluate the quality of leather. Good quality of leather is expected to have a minimum tensile strength - min. 20 N/mm.square. The value indicates poor quality of leather fibre weave.

9.8.12 Stitch-tear strength

Stitch tear strength is determined to assess the strength of the leather during stitching of leather components for making leather products. Stitch tear strength of the leather must be 100N/mm.

9.8.13 Abrasion resistance

It is carried out using Carborundum-coated paper as abrading material and is done using abrasion testing machine.

9.8.14 Elongation at break

It is the indication that leather can stretch or wear before it is broken. Elongation at break must be 45–85%.

9.8.15 Water vapour permeability

This property is important from comfort point of view, and is measured for upper, garment lining leather, etc by water vapour permeability apparatus: water vapour permeability (Herfeld) min.300 for upper leathers and 350 for garment leathers. It is a test to determine the amount of water vapour that a material will transmit through its structure in a specified time. This is an important performance measurement for leathers as it will determine the category/type of the leathers and subsequently the performance standards, it must need.

9.9   Performance standards of leather and suede garments[3]

9.9.1   USA market

Performance standards of leather and suede garments for USA market are given in Table 9.1.

9.9.2   EU market

Performance standards of leather and suede garments for EU market are given in Table 9.2.

9.10   Performance standards of leather and suede hides

Performance standards of leather and suede hides for USA market are given in Table 9.3.

9.11   Performance standard of leather garments for different buyers

9.11.1   Timberland[4]

Some of the leather testing which is followed by Timberland in addition to standard garment test is tabulated in Table 9.4.

9.11.2   Calvin Klein[5]

Some of the leather testing which is followed by Calvin Klein in addition to standard garment test is tabulated in Table 9.5

9.12   Concluding remarks

Leather goods, including clothing, belts, as well as small goods are not only high value items, but are also highly variable in terms of species, types of leather, dyeing/colour, chemicals employed in finishing, and microbiological activity. Therefore, it is important to identify the species and the chemicals used in dyeing and finishing, ensuring that the products meet quality requirements, and the safety requirements of the destination market The precision and

Table 9.1 Performance standards of leather and suede garments for USA market

Test Property	Test Method	Test Principle / Requirement
Labelling Guide for select leather and imitation leather products Textile Fiber Products Identification Act (if textile lining) Wool products Labelling Act (if wool lining) Care Labelling Rule	Visual Assessment	Leather Identification Fiber Content Label RN# or WPL# Country of Origin Care Label Size
Fiber Construction Defects	Visual Assessment	Report all defects
Dimensional Stability Professional Leather Clean	Professional Leather Clean	2% Shrinkage No growth Differential Shrinkage (Shell/Lining): 2%
Appearance after Cleaning General Appearance	Visual Assessment	No distortion, defects, damaged components, excessive puckering Pill – Class 4.0 Colour change: Class 4.0 Self staining: Class 4.5
Performance Component attachment strength	ASTM D 7142	15 lbs@10sec
Seam strength	ASTM D 1683	25 lb Min
Seam slippage	ASTM D 1683	20 lb Min @ ¼"
Analytical Lead in Surface Coating (When request)	ASTM E 1613 / 1645	Shall not exceed 90 ppm (0.009% by weight)

Table 9.2 Performance standards of leather and suede garments for EU market

Test Property	Test Method	Test Principle/ Requirement
Azo Colourants	EN 14362 – 1/2; ISO/TS 17234, In House Method	<30 mg/kg
Nickel (Qualitative)	Per EN 12471	Negative
Nickel Leaching (When request)	EN 12474/EN 1811	$<0.2 \ \mu g/cm^2/week$
Formaldehyde: Qualitative	AATCC 94	Negative
Formaldehyde: Quantitative (When request or nylon fiber included)	ISO 14184 – 1	<75 mg/kg
Sleepwear Flammability	Nightwear Regulation	Comply
Fiber Label verification	Determined by general garment testing procedures	Comply
Care Symbol wording verification	Determined by general garment testing procedures	Comply

Table 9.3 Performance standards of leather and suede hides for USA market

Test Property	Test Method	Test Principle/ Requirement
Fabric Construction Fiber analysis (Textile components / Lining)	AATCC 20 / 20A	Confirm
Leather thickness (millimeters)	ASTM D 1813	± 5 % from contracted weight
Defects	Visual Assessment	Report all defects
Dimensional Stability		
Professional Leather Clean	Professional Leather Clean	2% Shrinkage No growth

(Contd.)

Test Property	Test Method	Test Principle/ Requirement
Appearance **After Cleaning** General Appearance	Visual Assessment	No distortion, defects, damaged components, excessive puckering Pill – Class 4.0
Colour fastness CF to Professionally Leather Clean	Leather Clean – 1 Cycle	Colour staining: Class 4.0 Colour change: Class 4.0 Self- staining: Class 4.5
CF to Crocking	AATCC 8 / 116	Dry class: 3.0; Wet class: 4.0
CF to Light@20 AFU	AATCC 16, option 3	Class 4.0 (provide 10 AFU for reference when fails to 20 AFU
CF to Water: Colour block/contrast trims only	AATCC 107	Colour staining: Class 4.0 Colour change: Class 4.0 Self- staining: Class 4.5
Phenolic Yellowing: white, ivory if > 50% coverage	ISO 105 X18	Class 3.5
Performance Flammability – fabric only if not exempt	CFR 1610	Class 1
Seam slippage	ASTM D 434	20 lb Min @ ¼"
Tensile strength	ASTM D 5034	40 lb Min
Tear strength	ASTM D 1424	<6.0 oz/yd^2: 2.5 lb min $6.0 - 7.9$ oz/yd^2: 3.0 lb min 8.0 oz/yd^2: 4.0 lb min
Analytical Lead in surface coating (when request)	ASTM E 1613 / 1645	Shall not exceed 90 ppm (0.009% by weight)

Table 9.4 Performance standard of leather garments for timberland

Test	Testing method	Pigmented leather	Semi aniline leather	Aniline leather	Nubuck and Suede
Colour fastness to light	BS EN ISO 105-B02:1999	4	4	2/3	4
Colour fastness to rubbing	BS EN ISO 105 X12: 2002	Dry 3, Wet 2/3	Dry 3, Wet 2/3	Dry 3, Wet 2/3	Dry 3, Wet 2/3
Colour fastness to water spotting	BS EN ISO 15700:1999	3 (no residual halo)	3 (no residual halo)	3 (no residual halo)	3 (no residual halo)
Colour fastness to water	BS EN ISO 105 E01:1996	3	3	3	3
Colour fastness to drycleaning	BS EN ISO 105 D01:1995	3 (no finish loss, no reoiling)	3 (no finish loss, no reoiling)	3 (no finish loss, no reoiling)	3 (no finish loss, no reoiling)
Tear strength	BS EN 13937-1:2000	20 N	20 N	20 N	20 N
Water repellency	BS EN ISO 14340:2004	100	90	80	80

(Contd.)

Resistance to clod crack (applicable to leather with a coherent finish)	EN ISO 17233:2002	No finish cracks −10°C	No finish cracks -10°C	Not required	Not required
Finish adhesion (applicable to leather with a coherent finish)	BS EN ISO 11644:2003 (Dry Adhesion)	2N/10mm	2N/10mm	Not required	Not required
Flex resistance (applicable to leather with a coherent finish)	BS EN ISO 5402:2002	Not required	No finish cracks after 20,000 cycles	Not required	Not required

Table 9.5   Performance standard of leather garments for calvin klein

Example End use	Crocking dry/wet Minimum	Colorfastness to water Min. staining	Tensile Strength (Dumbbell Method) Minimum	Stitch Tear Strength (Double Hole) Minimum
Full Grain	3–4/3	3–4	2000 psi (no cracking)	10 lbs
Suede	3/2	3	2000 psi (no cracking)	10 lbs
Appearance	Color change: 3–4			

capabilities to validate professional leather clean labels not only reveal any deficiencies, but also can suggest remedial measures. Tests, such as drycleaning fastness, light fastness, finish adhesion, and veslic rubbing can be carried out on finished garment or materials and components prior to manufacturing. Components, such as zip, linings, interlinings, and buttons should also be assessed for performance. In addition, a cleansing assessment on the whole garment may also be considered.

References

1. Manual of leather garments (2015), Chapter 1: Introduction to Garment Leathers and Garments, 1–5.
2. Manual of leather garments (2015), Chapter 8: Quality control aspects, 104-106.
3. Casual Make Retail group, Inc. (CMRG) Textile Testing Manual (2011), Prepared by SGS, December 21, 59–60.
4. Timberland Apparel Vendor Manual (2011), Material and garment testing requirements, 10.
5. Calvin Klein Quality Manual (2012), Fabric and garment testing, Performance leather standards, 12.

CHAPTER 10

Role of different stakeholders in the quality management of apparel

Abstract

The understanding of the role of retailer, buying agent, factory, third party laboratory, and testing standard development organisation in the quality program of clothing sector is necessary to ensure the technical and commercial success of merchandise. The chapter first discusses importance of syncronisation of the activities of different stakeholders to deliver the right product within the stipulated time frame. The chapter then discusses the functional aspects of retailers, buying agents, factories, third party laboratories, and testing standard development organisation to engineer the apparel with compliance to quality standard.

Keywords: quality program, retailer, buying agent, factory, third party laboratory, AATCC, ASTM, ISO

Chapter contents (Section headings)

10.1 Introduction

Quality program in clothing sector involves various stages of operation. Each of them is essential to derive the commercial success of a particular garment style in focus. Conceptualisation of a product is of primary importance. The engineering of the garment[1] requires capability of a designated factory to produce the particular specification while taking care of the social compliance,[2] and technical excellence. In order to ensure those attributes, social compliance audit, and factory audit[3] are of paramount importance. Once an approved factory started to produce an ordered product, testing ensures assurance of its quality. A third party laboratory,[4] vendor, and the buying agent work closely to monitor the situation with a view to achieve high-quality merchandise. Inspection as per international or buyers' approved norms[5] is the next step to ascertain the quality of a garment. This consists of pre-production inspection, during production inspection and final random inspection. Quality evaluation at the production stage and at the pre-shipment stage is also performed by close co-ordination between factory, buying agent and third party nominated agency like SGS, BV and ITS—the world's leading inspection, verification, testing, and certification companies. Finally, when the quality of a particular product is conformed as per buyers' specification and a favourable rating is attained, shipment of merchandise is executed within stipulated time. This ensures that the consignment reaches to a particular country and is placed in the stores well ahead of the season to meet the expectations of the consumers. In this context, specialist apparel Warehouse Management System (WMS)[6] has already become a key factor in providing suppliers and their distribution partners with competitive advantage. Three parties in logistics mix must work closely together to enable quick response to demand, namely the retailer, the apparel supplier and in an increasing number of cases, a third party logistics provider. Supply chain management[7] will not have any major impact on costs of the product. These investments are being made by retailers to service customers better and offer them right goods at the right time. However, the chain of operation from the product design, manufacture, and quality assurance to the final shipment requires accurate harmonisation of activity between retailer, buying agent, vendor, and a third party consumer testing laboratory. In other words, ultimate commercial and technical success of an apparel product in international market depends on how best the coordination of different agencies was achieved.

In modern business, effective communication along the supply chain and with legislative bodies, clients, and customers is imperative. Standardisation can deliver measurable benefits when applied within the infrastructure of a company itself. Business costs and risks can be minimised, internal processes streamlined, and communication improved. Standardisation promotes interoperability, providing a competitive edge necessary for the effective worldwide trading of products and services. Quality standards are formulated by various international organisations, e.g., AATCC, ASTM, BSI, ISO, DIN, BIS, GB and others.

10.2   Role of a retailer[8]

1.   Work with lab to develop standards as per specific end use of the product and formulate a quality manual to register those materials as controlled copy document, which has to be used for any future requirement of quality assurance as the case may be.
2.   Evaluate the capability and suitability of a factory/supplier in terms of process, procedures, equipment, and general standards to produce a product that meets the requirements of their customers.
3.   Inform factories and agents of QA program and update them on test methods, sample submission requirements, and performance standards as changes are made for any apparel product.
4.   Advise factories/suppliers of the nearest approved testing laboratory for the submission of test requests.
5.   Review final reports of quality certification in testing and inspection and to follow-up on negative report results. If the submission has been rated as a "failure" or with "corrective action required", the findings are to be discussed internally and in partnership with the cross functional team followed through with a disposition to re-work, re-test, or cancel the order.
6.   Internal QA department provides risk assessment and advises their buyers appropriately.

10.3   Role of a buying agent[9]

1.   Inform factories of QA requirements of retailer.
2.   Evaluate the capability and suitability of a vendor/supplier in terms of process, procedures, equipment, and general standards to produce a product that meets the requirements of the retailers' customer.
3.   Assist with sample submission of the merchandise to the third party laboratory.
4.   Ensure vendor meets quality standards, delivery times and monitor the production to ensure consistency.
5.   Communicate the audit and test report findings to retailer.

10.4   Role of a factory[10]

1. All fabrics and trims must be sourced ensuring they comply with the minimum standards specified in QA specifications. Likewise all components such as sewing threads, zips, buttons, or tuck button, studs and so on must be sourced from approved or reputable sources and should meet the appropriate performance levels.
2. Must meet the legal requirements applicable to the country of production as well as code of conduct, social, environmental, and product quality standards. Such standards are not intended to restrict or hamper a vendor/suppliers capabilities or competitiveness.
3. Complete test request forms.
4. Send samples to the laboratory with relevant information such as garment description, style/PO, and the designated wash care instructions.
5. Send test reports to the buying agent and/or retailer.
6. Regardless of whether or not a third party laboratory has tested a particular item, the vendor/supplier is solely responsible for ensuring that their products meet or exceed all regulatory standards and voluntary industry standards in the markets in which they are sold. Products which fail to meet any such applicable local, state, or, the retailer will not accept national standards.

10.5   Role of a third party laboratory[11]

1. Maintain equipment as per a pre-determined schedule.
2. Monitor regulatory changes.
3. Conduct internal audits and correlations studies in proficiency testing[12].
4. Conduct testing and evaluate products in accordance with the quality standards of the retailer.
5. At the conclusion of testing, complete a test report. Where there is a "Failure" rating or a comment with "Corrective Action Required" is indicated, it is desired to include the reasons and recommendations for improvements for the items.
6. If a test request form is sent incomplete and/or insufficient sample is submitted to the lab, the third party nominated laboratory will contact the originating office. Until the testing laboratory receives proper information and samples, the testing will be put on hold, resulting in delays, which the vendor/supplier will be held responsible for.

7. Complete reports on timely basis.
8. Archive reports.
9. Preserve tested swatches for an agreed period of time.
10. Maintain laboratory accreditation status under ISO/IEC 17025:2017.

10.6 Role of a testing standard development organisation

1. The AATCC (American Association of Textile Chemists and Colorists) is internationally recognised for its standard methods of testing fibres and fabrics to measure and evaluate performance characteristics like colourfastness, appearance, soil release, dimensional change, and water resistance. They are published annually in the AATCC Technical Manual.[13]
2. ASTM International (formerly the American Society for Testing and Materials) is a globally recognised leader in the development and delivery of international voluntary consensus standards. It contains over 12,000 ASTM standards, published in 15 sections.[14]
3. The British Standards Institution[15] publishes a number of textile and clothing standards. The world's first management systems quality standard, BS 5750, was published by BSI in 1979. It produces standards and information products that promote and share best practice. Over 30,000 BSI standards and publications are created.
4. ISO (The International Organisation for Standardisation) publishes many textiles standards.[16] It is the world's largest developer of standards; their principal activity is to develop technical and economical standards. The work is normally carried out through ISO technical committees. In addition many European and Domestic versions are published as: EN ISO, European version of the International Standard and BS EN ISO, British version of the International Standard.
5. Bureau of Indian Standards[17] is a statutory body set up, established in 1986. The Bureau is a body corporate and responsible for formulating National Standards. It interests the field of standardisation, quality control, quality management system, environmental management system, laboratory management, etc.
6. Deutsches Institute fuer Normung (DIN, German Institute for Standardization) has been based in Berlin since 1917.[18] Its primary task is to work closely with its stakeholders to develop consensus-based standards that meet market requirements.

7. GB standards[19] are the China national standards, also called as Guobiao Standards, China GB standards are classified as two stages, Mandatory or Recommended. Mandatory standards have the force of law as do other technical regulations in China. They are enforced by laws and administrative regulations and concern the protection of human health, personal property and safety. All standards that fall outside of these characteristics are considered recommended standards. In China, China GB standards can be identified as Mandatory or Recommended by their prefix code. Prefix code GB are Mandatory standards and GB/T are recommended standards (Quasi-Mandatory standards). All products or services must be complied with GB standards, no matter domestic or imported products. Any products being sold in China are required to be tested in order to ensure their compliance with GB standards. If one wants to export products or services to huge Chinese market, need to understand and be aware of the complexities and necessary requirements under the vast range of GB standards, need to ensure they are meet the requirements of GB china national standards. The outcome of failing to comply with GB standards can include the rejection of products during importation as well as products being seized from stores, resulting in a significant impact on retailers and manufacturers in terms of reputation and cost.

10.7 Concluding remarks

The demand on the properties, appearance, and durability of materials and components in the apparel sector has increased significantly in recent years. Simultaneously, increasing competition has forced the industry to progressively reduce costs of end product. In order to meet these changing requirements, and to provide an objective framework for what is acceptable in export to different destinations, quality characterisation of apparels has continuously been attracted attention. It is definitely essential in the perspective of ensuring the right quality, confidence of wearing and protecting the health and safety aspects. Well-known brands in the world normally accept the merchandise based on the conformity as per their quality benchmark. Various regulations and standards vary depending upon the country. But ultimate success of the quality of merchandise depends on the right coordination between the retailer, buying agent, vendor, the third party laboratory, and the testing standard development organisation.

References

1. Dragcevic Zvenko, Zavec Daniela, Rogale Dubravco, and Gersak Jelka (2002), 'Workloads and strandard time norms in garment engineering', *J Text and Apparel, Tech and Mgmt*, 2, 1.

2. Chaterjee Abira (2008), 'Social compliance, social accountability and corporate social responsibility', *Mainstream*, 156, 18.

3. Murphy F David and Mathew David (2001), 'A case study prepared for the new academy of business innovation network for socially responsible business', *Nike and global labour practices*, 1–32.

4. Das Subrata (2008), 'Strategy to leverage KAM clients in third party services to apparel sector'. Available from: www.fibre2fashion.com [Accessed 6 March 2009].

5. Wilsey Ken (2001), 'Inspection requirements and procedures', Supreme international quality assurance manual, USA.

6. ARC advisory group (2008), 'Warehouse management systems'. Available from: http://www.arcweb.com/Research/Studies/Pages/WMS.aspx [Accessed 5 March 2009].

7. Cooper M C and Ellram L M (1993), 'Characteristics of supply chain management and the Implications for purchasing and logistics strategy', *The Intl J of Logistics Mgmt*, 4, 13–24.

8. Kmart far east procedures manual for softline testing program of apparel & apparel accessories (2000), Kmart merchandising department responsibility, Section IV, December.

9. Russell corporation quality assurance (2001), Testing requirements and procedures, Buyer responsibility, USA.

10. Kmart far east procedures manual for softline testing program of apparel & apparel accessories (2000), Kmart vendor responsibility, Section VI, December.

11. Kmart far east procedures manual for softline testing program of apparel & apparel accessories (2000), Kmart appointed laboratory responsibility, Section VII, December.

12. ISO/IEC Guide 43-1:1997, Proficiency testing by interlaboratory comparisons, Part1: Development and operation of proficiency testing schemes.

13. AATCC, 03018A: 2018, AATCC Technical Manual, USA.

14. ASTM International, Standards and Publications, USA. Available from: https://www. astm.org/Standard/standards-and-publications.html [Accessed on 2 April 2018}

15. British Standard Institutions, United Kingdom. Available from: https://www.iso. org/standards.html_[Accessed on 2 April 2018].

16. International Organisation for Standardisation, Available from: https://www.iso. org/news/ref2250.html [Accessed on 2 April 2018].

17. Bureau of Indian Standards, The National Standards Body of India. Available from: http://www.bis.gov.in/ [Accessed on 2 April 2018].

18. Mit Normen und Standards den digitalen Wandel gestalten, Hannover Messe 2018. Available from: https://www.din.de/de/din-und-seine-partner/termine/mit-normen-und-standards-den-digitalen-wandel-gestalten-264710 [Accessed on 2 April 2018].

19. China National Standards, GB Standards. Available from: http://www.gbstandards.org/ [Accessed on 2 April 2018].

Index